Property Asset Management

OTHER TITLES FROM E & FN SPON

Caring for our Built Heritage
A survey of conservation schemes carried out by County Councils in
England and Wales
T. Haskell

Competitive Cities
Succeeding in the global economy
Hazel Duffy

Industrial Property Markets in Western Europe
Edited by B. Wood and R. Williams

Land and Property Development
New directions
Edited by R. Grover

Marketing the City
The role of flagship developments in urban regeneration
Hedley Smyth

Microcomputers in Property
A surveyor's guide to Lotus 1-2-3 and dBase IV
T.J. Dixon, O. Bevan and S. Hargitay

The Multilingual Dictionary of Real Estate Terms
A guide for the property professional in the single European Market
Edited by L. van Breugel, B. Wood and R. Williams

National Taxation for Property Management and Valuation
A. MacLeary

Property Development
Third Edition
D. Cadman and L. Austin-Crowe

Property Investment and the Capital Markets
G.R. Brown

Property Investment Decisions
A quantitative approach
S. Hargitay and M. Yu

Property Investment Theory
Edited by A. MacLeary and N. Nanthakumaran

Property Valuation
The five methods
D. Scarrett

Rebuilding the City
Property-led urban regeneration
Edited by P. Healey, D. Usher, S. Davoudi, S. Tavsanoglu and M. O'Toole

Risk Analysis in Project Management
J. Raftery

Urban Regeneration, Property Investment and Development
J. Berry, W. Deddis and W. McGreal

UK Directory of Property Developers, Investors and Financiers, 1993
Building Economics Bureau

Effective Speaking
Communicating in speech
C. Turk

Effective Writing
Improving scientific, technical and business communication
2nd Edition
C. Turk and J. Kirkman

Good Style
Writing for science and technology
J. Kirkman

Write in Style
A guide to good English
R. Palmer

Journal

Journal of Property Research
(Formerly Land Development Studies)
Edited by Bryan D. MacGregor (UK), David Hartzell and Mike Miles (USA)

For more information on these and other titles please contact:
The Promotion Department, E & FN Spon, 2–6 Boundary Row, London SE1 8HN.
Telephone 0171-522 9966

Property Asset Management

Second edition

Douglas Scarrett

E & FN SPON
An Imprint of Chapman & Hall

London · Glasgow · Weinheim · New York · Tokyo · Melbourne · Madras

Published by E & FN Spon, an imprint of Chapman and Hall,
2–6 Boundary Row, London SE1 8HN, UK

Chapman & Hall, 2–6 Boundary Row, London SE1 8HN, UK

Blackie Academic & Professional, Wester Cleddens Road, Bishopbriggs, Glasgow G64 2NZ, UK

Chapman & Hall GmbH, Pappelallee 3, 69469 Weinheim, Germany

Chapman & Hall USA, 115 Fifth Avenue, New York, NY 10003, USA

Chapman & Hall Japan, ITP-Japa, Kyowa Building, 3F, 2-2-1 Hirakawacho, Chiyoda-ku, Tokyo 102, Japan

Chapman & Hall Australia, 102 Dodds Street, South Melbourne, Victoria 3205, Australia

Chapman & Hall India, R. Seshadri, 32 Second Main Road, CIT East, Madras 600 035, India

First published in 1983 as *Property Management* by E & FN Spon
Reprinted 1991
Second edition 1995

© 1983, 1995 Douglas Scarrett

Typeset in $10\frac{1}{2}/12\frac{1}{2}$ Times by Photoprint, Torquay, S. Devon
Printed in Great Britain by The Alden Press, Osney Mead, Oxford

ISBN 0 419 19310 3

A catalogue record for this book is available from the British Library

Library of Congress Catalog Card Number: 95–68509

∞ Printed on permanent acid-free text paper, manufactured in accordance with ANSI/NISO Z39.48-1992 and ANSI/NISO Z39.48-1984 (Permanence of Paper).

Contents

Preface

The first edition set out to provide a practical guide to property management. This new edition retains the practical approach which has proved popular with students and practitioners alike. At the same time, the text recognises and reflects changes in the approach to property management by investors and users generally.

The wide use of computers for records and accounting has released managers from the necessary but previously time-consuming aspects of management and enabled them to contemplate and concentrate on the strategic considerations necessary to optimise the asset, whether held as an investment or for operational use within a firm or company. In times of recession, both landlords and tenants may be more conscious of the contractual arrangements between them. Some of that awareness is evident in the high level of litigation, for example on the interpretation of rent review provisions and the liability of sureties and guarantors. A representative number of cases is reported briefly, but the reader is advised to study the full report in each case to gain the full details.

I am grateful to colleagues and friends for advice and assistance generally and in particular to Stuart Planner for his considerable help in organising the disks for submission to the publisher.

Table of cases

Table of statutes

<table>
<tr><td colspan="2">

The role and responsibilities of the property manager

</td><td>

1

</td></tr>
</table>

This chapter introduces the uniqueness of each property interest and emphasises the difference between property and other forms of investment media, not least because of the nature of the relationship between landlord and tenant. It is emphasised that quality of management is an important contributor to performance: in this context the duties of managing agents are outlined as are their legal responsibilities.

1.1 PROPERTY OWNERSHIP

Ownership of property – land and buildings – with the express intention of letting to tenants, usually on fairly long-term contracts, is a well-established investment medium as an alternative or supplementary to stocks (being investments in government-backed borrowings referred to as 'gilt-edged' securities) and shares (being investments in the equity market where the return relies on the performance and policy of the company in which the shares are held). Properties are also acquired for operational purposes. The range of investment opportunities is shown in Figure 1.1.

1.2 LAND AS A RESOURCE

Of the principal factors of production – workforce, raw materials, machines, energy and property (land and buildings, also referred to as the fifth strategic resource in production decision theory) – the operational role of property is not well understood. The problem was first highlighted by a succession of reports by the National Audit Office and the Audit Commission for Local Authorities in England and Wales which exposed deficiencies in the management of property resources in the public sector and suggested ways in which greater efficiency could be achieved. The Audit Commission, for example, published *A Management Overview* and *A Management Handbook* together dealing with strategic issues and good practice in the management of local authority property (excluding housing). Arising from this initiative, local authorities now have a heightened awareness of the place of property and the need for total occupation costs and other information on which future decisions can be based. A similar initiative was undertaken at Reading University where a team of researchers carried out a detailed study of operational property in use by

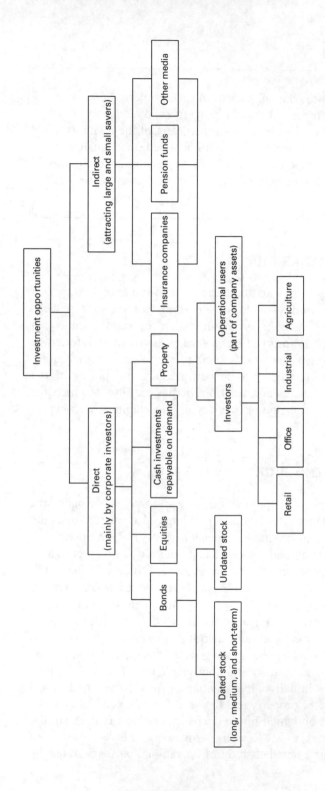

Figure 1.1 Range of investment opportunities.

230 of the largest public limited companies in the United Kingdom to find that typically, although the company's property holding accounted for some 30% of its total asset value, it was often managed in a reactive way if at all and with incomplete records.

These enquiries and the wide dissemination of their findings, together with the general state of business in recent years and the pressure on local authorities to cut costs have encouraged more thought on the part of both owners and occupiers of commercial buildings.

The receipt of a notice to quit, designed as a first stage in the grant of a new lease and only intended by the landlord as an opportunity to increase the rent, may now be the catalyst for a review both of the use of the premises and the organisation of the business in the course of considering whether the company wishes or needs to and if so can afford to continue to occupy the premises.

1.2.1 Property as an investment

There is no prospect of value growth in building society investments or many government securities. Equities and some gilts offer this combination, although the range and diversity makes comparison more difficult. Investors ideally aim for a positive or real interest annual return, an element of capital appreciation and a reasonable level of security. Yet in order to maximise returns it is essential to compare the total likely out-turn on the investment.

The effect of capital change is less certain since it exists only on paper until the asset is sold. It is this aspect more than any other that sets property apart from the majority of other investments and is referred to as illiquidity. Investors in gilts and equities can make a judgement about the current value of their holdings and decide to sell in the knowledge that the transaction will take place without delay and on the terms agreed. The delay in implementing any property investment decision and the likelihood of market changes before it is effected tends to place property apart from other principal forms of investment. The investor will expect the relative illiquidity to carry an uncertainty premium as a trade-off.

1.2.2 Demand for land and buildings

Banks, insurance companies, building societies and many of the leading retailers own their own operational premises where possible as a matter of policy. Manufacturers in the basic industries, especially where specialised processes use few conventional buildings, often have no option but to provide their own premises, developers not being prepared to finance such developments which often have no alternative use, suffer heavy wear and tear and are susceptible to technological change with a direct but uncertain effect on the remaining useful life of the unit.

Many of the remaining retailers, office users and manufacturers take a conscious decision to lease their premises, enabling available capital to be employed in the furtherance of their core business activities. The security offered to tenants by legislation provides real choice and has done much to promote the acceptability and popularity of leasehold interests among commercial users. At the same time, it restricts their exposure to risk of change and maintains their flexibility to relocate or relinquish if necessary for any reason.

Fashions and preferences change. Some of the largest retail chains which command sufficient capital to enable them to own many of their operational properties in preference to renting, have opted to become leaseholders, at least in some of their premises, releasing capital by selling blocks of property to property companies and others with the benefit of their covenant as the basis of a lease back. Recent examples include two of the more successful retailing groups, Dixons and Sainsbury's. Companies appreciate the option; relatively new and growing businesses will tend to look upon leased premises with favour as enabling them to expand at a faster rate than would be the case with freehold acquisitions.

The provision of premises for letting to these various users has become a major industry, almost entirely controlled by property companies, insurance companies, pension funds and similar institutions, which have vast investments in property. The majority of investments by such groups is in commercial (retail and offices) and industrial premises (factories, warehouses) with relatively small holdings in agricultural land and possibly residential property.

1.2.3   The available market

An important aspect of investment is diversification which may be achieved by investing across sectors and in several geographical areas. For the larger funds, the marketplace is global: even where there is some hesitation about investments around the world, there are many opportunities in Europe.

The wellbeing of the property market is inextricably linked to the economic performance of the country and its success in the European dimension. The deep structural changes of the 1980s are still working their way through the system and whichever political party is in power the result is likely to be further change and uncertainty. There is an urgent need to operate within the European community in a way which will ensure substantial benefits from the larger market to at least justify Britain's substantial and increasing financial contribution. Property, whether held as an investment or used operationally, has an important part to play in this integration.

1.3 THE PROVIDERS OF LAND AND BUILDINGS

Property may be provided for the use of others by an investor; by a company for use in its business and to enable their operations to function and grow; and by various public bodies, local and national, mainly for their own operational use but including an element of investment property.

It is not possible to assess the capital value of property holdings with any precision although the result of a careful enquiry in 1989 put the value of UK holdings of commercial property at £250bn of which some 7% was held by the eight largest property companies. At about the same time another source placed the operational assets of the 29 largest United Kingdom companies at £58bn.

Almost all the significant holdings of property are held corporately, by pension funds, institutions, property companies and business firms, hospital authorities and similar groups.

Individuals usually invest in stocks and shares and indirectly in property through premium payments to insurance companies and contributions to private or company pension funds. These commercial organisations together with public and quasi-public undertakings have vast investment holdings. The quasi public ones include the Crown estate, the Civil estate, the Church commissioners and the Duchy of Cornwall.

The considerations of management will be determined to a large extent by the declared purpose of the organisation and the need to satisfy various interested parties – shareholders, policy holders, debenture holders, pension recipients, the Registrar of Friendly Societies, the Charity Commissioners among others – that the investment is being managed in an appropriate way.

1.3.1 The main categories of investors in property

Large and small contributions from individual and corporate sources provide the funds necessary for investment in the various investment media. They may be deposited by individuals or their employers to provide pensions at some future date or to finance the redemption of a mortgage, personal savings, regular or spasmodic and often of relatively modest amounts, or by professional investors in larger amounts.

(a) The private sector

By far the largest concentrations of property are held corporately by pension funds, property companies and business firms.

Three examples give an example of the extent and spread of the holdings of larger corporate investors:

1. One of the larger United Kingdom insurance funds is Scottish Widows which has investments in its Managed Pension Fund and Unit Linked Fund with a total value of some £1.25bn in the year ending March 1993. Approximately 7% of its life fund is invested in the property sector, principally in retail and industrial properties. Most management is carried on in-house although agents are engaged to search for and acquire suitable investment properties.
2. The Liverpool Victoria Friendly Society has some £3.2bn invested assets, including £446m mainly in commercial property. A good deal of restructuring has taken place in the last few years to rebalance the portfolio and to improve the quality and lot size of the stock.
3. The Grosvenor Estate is the largest privately owned investment fund in the United Kingdom. The nucleus of the estate has been in the family ownership for over 500 years, based on extensive land holdings in London and Chester. Fortuitously, the land in London was and remains in areas favoured for high-quality housing and premier shopping areas. The trust has a sizeable subsidiary operation where substantial and prestigious assets are controlled.

(b) The public sector

Central government owns much of its older operational stock, mostly in the capital but with other substantial holdings, such as Crown post offices, in towns and cities throughout the country. There has been a tendency in recent times to enter the market and take leases of office blocks and other types of property.

In the past, local authorities have acquired vast amounts of land within their own area. For example, it has been estimated that Manchester and Sheffield each own approaching 50% of the land within their respective areas. The property holdings of English and Welsh local authorities were estimated to have a capital value of some £100bn in 1989 (excluding residential housing stocks). More surprisingly, it has been estimated that, of this total, at least 25% is regarded as investment holdings (although the use of many such investments may be tempered and the return possibly curtailed by the social aspirations of the authority).

A sizeable part of the housing stock is held by local authorities and housing associations, whilst private landlords have a smaller holding and public and private bodies maintain housing stocks as an aid to the furtherance of their principal activity (for example, student bedspaces owned by universities or hospitals).

The local authority stock of housing to let is ageing, with few new developments having been completed in recent years: it has been much depleted by mandatory sales to those tenants qualifying and wishing to buy at a discount. At the same time housing associations have become an important force in the market for renting and part ownership schemes, having received finance from government in recent years in preference to local authorities. They are now expected to finance new development by raising at least part of the capital required in the commercial market.

(c) Quasi-public bodies

The Crown estate commissioners for example, control rents and profits traded by George III in 1760 in exchange for the greater certainty of the civil list with a capital value in 1992 of some £1.6bn and a revenue surplus of £74m and the Church of England (capital value of £1.4bn in 1989 but much eroded by 1992 as the result of heavy losses on certain investments, having serious implications for future clergy stipends).

A number of old established and substantial trusts include important property holdings in their assets. Here are some examples:

- The Crown Estate dates back to the reign of Edward the Confessor; in 1760 George III traded annual rents and profits to Parliament in exchange for the greater certainty of the Civil List. Apart from important commercial holdings in shops and offices, mainly in the city of London but increasingly elsewhere in the country, there is a small amount of housing and over 300 000 acres of agricultural land, fish farms in Scotland and the Crown's rights to half of the nation's foreshore and an extensive offshore sea bed. The most recent accounts (to March 1994) show a total portfolio with a value in excess of £2bn,

75% of which is contained in urban areas. In the same year, a surplus of £79m was reported. There was also a modest investment in the same year of some £40m in development projects, some undertaken in partnership. The estate has no borrowing powers. Measured by capital value, some 27% of the estate is managed in-house, the remainder being contracted out to four main agents. Total payments of fees to agents during the year amounted to £5.5m. Until recently, the same firms have enjoyed security and the management has benefited from continuity but the new policy is to offer management contracts for competitive tender or to otherwise carry out market testing.

- The Duchy of Cornwall is a major estate mainly located in the west country, providing much of the income received by the Prince of Wales. The portfolio includes business and residential tenants, some 120 000 acres of agricultural land, the Isles of Scilly and some specialist holdings such as the Oval cricket ground and oyster beds on the river Helford. It was originally endowed in 1337 for the support of Edward III's eldest son. Purchases, sales and leases are controlled by Act of Parliament with leases, except building leases being restricted to terms not exceeding 31 years. The Duchy is forbidden to lend or borrow for any purpose. Much of the estate is directed, controlled and managed by land stewards reporting back to the governing Council of the Duchy under the chairmanship of the Prince of Wales.

- The Church Commissioners control a major investment portfolio, the income from which is earmarked for church expenses including an increasing requirement for clergy pensions. The accounts for 1993 show total assets valued at some £2.64bn of which slightly less than half is held in property. Commercial property accounts for assets valued at £872m and contributed an income of £59 m in the year. The property portfolio includes offices, shops, warehouses, factories, farms and residential properties which have accounted for considerable sales in recent years. The largest single asset is also one of the largest shopping complexes in the country, the Metrocentre in Gateshead. It is noted that the proportion of assets allocated to property is far in excess of that normally committed by a pension fund; that, and the tremendous appetite for development property in the early and mid 1980s which accounted for expenditure of almost £1bn, left the fund in a difficult and exposed position as the recession started to take hold in 1989. The Lambeth group was set up to deal with the problems, the remedies include a change of professional advisers and the commissioning of a strategic plan for the property element of the fund as the start to the

Commissioners rebasing their operation in pursuit of a healthy future.

(d) Operational property

Very little regard has been paid in the past to the role of the tenant or occupier who uses the building for operational purposes and incurs a cost in purchasing or an ongoing cost by way of rent in anticipation of making a profit from the activity carried on – production or service provision – within the building. Where users own their premises, the value to them is their function as an integral part of their business operations. In normal circumstances, they would not regard it as an asset to be traded; any sale would require a replacement unless the building had become redundant. An exception would arise where the redevelopment value of the site was substantially above the value of the building and its replacement cost.

Many commentators regard the interests of landlord and tenant as inimical and incompatible but a moment's thought and reflection will suggest a complementary relationship. A trader might be very anxious to occupy a particular building to further his business ambitions. That does not mean that he will take a tenancy however unreasonable the rental or other terms may be. He will expect to be able to incorporate the effect of the terms in a business plan which leaves a return for the time and expertise they are proposing to invest in the new or extended activity. If he is able to achieve his objectives and be successful, the landlord has greater security of income and continuity of occupier. If the occupier is unable to reconcile the terms he will either abort the proposal or look for other ways of ensuring its fulfilment. If that involves him in funding the purchase of the property he will need to borrow or to find the funds from his own resources. In the latter case and unless his is a very large and established successful company, the need to divert capital is likely to slow down the rate of expansion. At this level, the acquisition of property may be seen simply as the cost of working capital though in reality it is much more involved than that.

Whether owner or tenant, the operational user can derive considerable advantage from the advisory services of a property manager.

There was a perception in some parts of the public sector that there was no cost involved in the provision of land and buildings simply because they were provided either in the form of capital grants or refund of loan charges by central government, and no annual charge was debited to the occupier. There was widespread misuse or underuse of buildings which, whilst not held for investment purposes, tied up capital and called on resources for running costs, maintenance and repair.

1.4 THE MANAGEMENT TASK

Much of this book concerns the optimisation of the owner's investment: the process so described is property management whether they are an investor seeking a financial return or an operational user, aiming for efficiency in all parts of their business. Each has broadly similar objectives, even though the route and the emphases may be different.

Sound management and the identification and exploration of potential are essential to the wellbeing of a commercially-orientated company and corporate ownership is more likely to recognise and exploit opportunities afforded by the longer-term horizons.

1.4.1 The attributes and organisation of sound management

Successful property management is a demanding activity requiring understanding, ability and technical and organisational skills and resources.

The manager must therefore have the ability to select, direct, meld and inspire a team with a range of talents so as to release them in the most potent way to the benefit of the client. Pure organisational ability is therefore important with the highest level of service being delivered at an acceptable cost. There is also a need for a clear appreciation of the client's aspirations on the one hand and the means of achieving them on the other.

Clients will determine the nature and extent of the services required. They may elect to make use of the full range of services available, as and when appropriate, or they may limit their contract to certain aspects of the full service. Any limitation in their requirements may reflect their own contribution and expertise or their view of the need for and appropriate provider of a particular aspect of the service.

The manager is the head of a group of people having a spread of interpersonal, professional and technical abilities deployed in ensuring the most efficient management possible. The team will be concerned with all those matters associated with the owner's objectives. In general the manager will seek to achieve a streamlined and efficient management service, continuing upgrading of a portfolio to secure a balanced mix, safe growth and containment of risk. Modern property management requires the contributions of several specialisms but general practice surveyors not only occupy a dominant role but traditionally also lead the team. They may be employed in public service or in private practice; they may be in full-time salaried employ of an institution or they may act as agents of their several principals on the basis of a negotiated fee or may charge a commission on the amount of rent collected. This was once the usual method but today, with greater emphasis on overall management, such a basis is not altogether appropriate. In-house managers often supplement their skills and abilities by the employment of private practitioners for special tasks or for certain limited management functions or tasks. For example, there may be a need for technical advice on a troublesome maintenance problem outside the competence of the regular team, or a need for a particular task to be completed to a very tight timetable. In such cases, engaging specialists will not only ensure that the best team does the job, it will ensure that the regular team is not diverted from its main responsibilities.

1.4.2 Management: a definition and some models

Sound management is an essential prerequisite of a healthy and success-ful investment, whether comprising one or two single properties, a substantial development or a mature, balanced portfolio. Management may be defined as the process of planning business objectives, policies and activities, harnessing resources to ensure implementation and exer-cising control towards achieving the objectives. Although it has been shown that the elements of organisation, decision making and policy implementation are all present, property management is sufficiently distinct as a discipline to require a more detailed and specific attempt at a definition.

Property management seeks to advise on the establishment of an appropriate framework within which to oversee property holdings to achieve the agreed short- and long-term objectives of the estate owner and particularly to have regard to the purpose for which the estate is held. The basic needs will be to carry out such tasks as negotiating lettings on suitable terms, initiating and negotiating rent reviews and lease renewals, overseeing physical maintenance and the enforcement of lease covenants. These activities will take place within an agreed strategic framework where there is a need to be mindful of the necessity of upgrading and merging interests where possible, recognising other opportunities for the development of potential and fulfilling the owner's legal and social duties to the community.

Today, the management of a cohesive group of buildings let to occupiers on the best terms available, where the primary purpose is to maximise financial returns to the benefit of the owner, is termed portfolio management, whereas the management of those buildings held for operational purposes is termed property asset management. It is suggested that the term asset management could be more widely used, in that all properties are assets whereas not all collections of properties are properly described as portfolios. The original meaning of the word was a neutral one referring to a bundle of papers, but it has been developed to refer to a collection of properties assembled with some skill to achieve a balance of risk and return and managed towards perfecting that balance. In a diversified investment portfolio, property will comprise only part of the total funds which will have investments in bonds and equities also, helping to move towards a perfect portfolio.

Management must be based on the terms of the contract between the parties on the one hand and on an appreciation and interpretation of the particular owner's objectives on the other.

1.4.3 Approaches to management

Individual funds own tens of hundreds of millions of pounds worth of property investments (for example, Prudential show the value of their property portfolios in various funds as £4.5bn). The manager with overall control of such a portfolio is responsible for operating a sizeable business venture, where the quality of the decisions made will be a critical factor in the overall success of the fund. There are four broad models of management in common use today:

(a) In-house management

The advantage of in-house management is that the team can be focused solely on the interests of that particular fund or company and develop a fast track response to investment decisions where appropriate. Property companies and the larger funds are in a position to act in this way, benefit from their particular skills and expertise and minimise the knowledge of the outside world of their activities. One limitation is that the group is unlikely to have the expertise on every facet of property and it would be uneconomic to take on highly-paid executives to deal with specialised areas unlikely to crop up frequently.

(b) Management by an appointed agent

The appointment of an agent to advise and manage is an option widely used and in many cases to good effect. Some of the larger agents have very substantial capital values under management. The advantages are considerable. The firms engage in a wide range of activities related to property and maintain highly experienced and motivated teams. It is possible to call on knowledge and expertise beyond the immediate management area; the marketing teams will have current experience of demand and rental levels and the valuation team will know what yields are being achieved in a variety of transactions. There are disadvantages. Whilst there is no suggestion that any impropriety would occur – most firms go to considerable lengths to ensure confidentiality – there is no exclusivity of engagement and the commitment is to a number of clients, rather than to one client (which is, of course, a strength also). The cost may or may not be more than that of an in-house team. In recent years, firms of agents have become used to competing for instructions on the basis of 'beauty contests', elaborate competitive presentations where the service is proposed, the cost or its method of determination and the impression made by the team all play a part in the eventual choice. Many such appointments are made for periods of three to five years before the appointment is reconsidered. Whilst the earlier practice of appointing an agent on a more or less permanent basis no doubt lead to some complacency and a lack of keenness on the part of a few agents, frequent changes of agent are hardly likely to foster commitment or enable the team to build up a background knowledge and ethos.

There are occasions where the owner of a large portfolio will engage different firms to act in different geographical locations or in different sectors of the property market. Sometimes, the agents themselves will appoint a local agent, with the approval of their principal and usually only to deal with routine and day to day matters.

(c) A combination of the first two

A combination of the two types described above can take many forms, whereby the in-house management and the agent have a respect for each other's abilities and expertise and work together on the range of management tasks in an almost seamless way to the benefit of the portfolio. It requires a particular kind of chemistry to enable such a partnership to be successful; it envisages two managers and two teams working together at various levels and intensities without either manager having total line management. Such an arrangement needs strong commitment from each team member if it is to succeed.

(d) A hierarchical division, whereby the in-house manager directs the strategic thrust, limiting the agents to carry out a limited management role

Many companies will opt for this model, where the thrust of the business is conducted by the in-house managers, leaving the appointed agent to carry out the basic and repetitive tasks of property management. There should be a contract between the parties setting out the precise extent of the agent's responsibilities so as to avoid misunderstandings. Agents may well find that a limited relationship of this kind will develop over the years and that more work will be entrusted to them, either as an extension to their list of duties or as one-off tasks. The reason for giving special assignments to an agent may be that in-house expertise is lacking or too limited, that a second opinion is sought or that the requirements coupled with the time scale are too great or too disruptive of other activities for the in-house team.

1.5 NEGLIGENCE AND THE PROPERTY MANAGER: GENERAL PROPOSITION

The person offering a professional service – doctor, architect, engineer, accountant, solicitor, property manager – hold themselves out as possessed of the competence inferred by their calling. Anyone falling short of a certain standard is liable to be sued for those shortcomings where damage is suffered as a consequence. There is one exception: for historical and judicial reasons a barrister cannot be sued for negligence although, following a report on standards, a Barristers' Complaints Bureau is proposed and it is anticipated that limited compensation for unsatisfactory service will be available from January 1996.

The general proposition was set out in a judgement delivered more than 150 years ago:

> Every person who enters into a learned profession undertakes to bring a degree of care and skill. He does not undertake, if he is an attorney, that at all events you shall gain your case, nor does a surgeon undertake that he will perform a cure; nor does he undertake to use the highest possible degree of skill. There may be persons who have higher education and greater advantages than he

has, but he undertakes to bring a fair, reasonable and competent degree of skill.

Lanphier v. *Phipos* (1838)

In a medical negligence action against two consultants, the claim failed, a general statement in an earlier case was quoted with approval:

In the realm of diagnosis and treatment there is ample scope for genuine difference of opinion and one man clearly is not negligent merely because his conclusion differs from that of other professional men ... The true test for establishing negligence in diagnosis or treatment on the part of a doctor is whether he has been proved guilty of such failure as no doctor of ordinary skill would be guilty if acting with ordinary care.

Hunter v. *Hanley* (1955)

The third edition of Halsbury's Laws of England considers the question of negligence in relation to work carried out by an architect although it is clear that the principles evinced apply to members of other professions also.

The question whether the architect ... has used a reasonable and proper amount of care and skill is one of fact, and it appears to rest on the consideration whether other persons exercising the same profession, and being men of experience and skill therein, would or would not have acted in the same way as the architect in question. It is evidence of ignorance and unskilfulness in any particular to act contrary to the established principles of art or science which are universally recognised by members of the profession.

In certain cases, professional people may make themselves liable in negligence to third parties in which case the action is pursued in tort.

1.5.1 Level of knowledge inferred

Some professionals – particularly architects, accountants, engineers and property managers – need some knowledge of the law relating to matters within their technical competence. In general terms this is unlikely to infer knowledge to the level expected of a solicitor: on the other hand there is so much technical law relating to, for example, planning, building regulations, the landlord and tenant relationship, taxation and compulsory purchase, that a lack of understanding in these areas could well be held to be negligent. This is illustrated by the case of *Weedon* v. *Hindwood*

Clarke & Esplin (1974) where a surveyor instructed to negotiate compensation in a compulsory purchase case failed to adopt the advantageous later date of valuation in computing the amount of compensation to be claimed, as established by the case of *West Midlands Baptist (Trust) Association (Inc.)* v. *Birmingham City Council* (1970). The decision in this case established that the date for assessment of compensation was the earliest date at which the claimant could reasonably start work of replacement, rather than the date of notice to treat, as previously believed.

1.5.2   Duty of care

The relationship between client and professional adviser is based upon the general precept that the individual adviser or firm enjoys the confidence of clients and accepts a duty of care to them as an inherent part of the contract.

That duty is one against the occurrence of which advisers are likely to insure in order to protect themselves and, in the case of most professions, one where their professional association requires them to carry adequate indemnity cover. Most prudent property managers will wish to restrict their liability not only to matters within their competence, but also within an agreed statement of the service to be provided and paid for as contained within a management contract. Certain exclusion clauses may be stipulated to draw attention to the limitations of the service provided, tending to emphasise potentially fertile areas of dispute and or misunderstanding.

Should performance fall short in any material respect there is a prima facie right to sue on the contract for negligence; third parties without any contractual relationship may be able to sue in tort whilst in some cases a claimant may be able to choose whether to proceed by way of an action in contract or tort, the advantage of the latter being a longer period under the Limitation Acts in which legal action may be commenced.

1.5.3   Negligence claims

The main publicity of negligence claims in the surveying profession has been in respect of valuations and structural surveys. The detailed nature of property management and the possibility of neglecting to take some action or of meeting statutory or contractual deadlines resulting in substantial financial loss has made the threat of legal action very real. As a result, case law has grown considerably over the last few years.

The duty of care is one shared with anyone accepting responsibility under a contract but tends to be more developed and pursued in respect of members of professions offering a specific service where it is alleged that an acceptable level of competence has not been achieved. The plaintiff needs to prove the existence of a duty of care arising from an express or implied contractual term or in tort.

It is not possible to sue arbitrators for the reason that they occupy a quasi-judicial role. Professional advisors incur liability as do the independent valuers or experts appointed to resolve the dispute between the parties.

The particular requirement is to exercise such care and skill as may reasonably be expected of a practitioner in all the circumstances of the case.

Mr Justice McNair said:

> The test is the standard of the ordinary skilled man exercising and professing to have that special skill. A man need not possess the highest expert skill, it is well established law that it is sufficient if he exercises the ordinary skill of an ordinary competent man exercising that particular art.
>
> *Bolam* v. *Friern Hospital Management Committee* (1957)

The standard is an objective one unaffected by the attributes of the individual practitioner. Thus, a trainee's lack of experience was of no avail in defending an action for negligence where in her naiveté she had handed keys to a confidence trickster with unfortunate consequences (*Brutton* v. *Alfred Savill, Curtis & Henson* (1971)).

1.5.4 Some examples of negligence

As an indication of the extent of the liability of property managers in the pursuit of their duties, the outcomes of some relevant cases of negligence are set out.

In *Nahhas* v. *Pier House Management Ltd* (1984) it was held that managers of a block of flats responsible for engaging staff who failed to investigate the past of a man employed as a porter who used keys in his charge to steal jewellery from one of the flats were liable to the tenant in negligence.

Negligent conduct was held to have occurred in a case where a firm of property managers failed to verify references of prospective tenants, to check the inventory, to obtain rent in advance and to ensure payment of bills, etc. (*Murray* v. *Sturgis* (1981)).

In a recent case, it was held that a receiver had power to trigger rent reviews and should have done so. He was not just a rent collector as he had claimed but had a general duty to look after the property in a wider sense and was negligent in not having done so (*Knight* v. *Lawrence* (1991)).

Where solicitors failed to apply to court for a new tenancy under the Landlord and Tenant Act 1954 their clients were awarded damages in the sum of £56 000 representing the additional value of a protected tenancy which they were denied by the omission of the solicitors (*Hodge* v. *Clifford Cowling & Co.* (1990)).

In a valuation for reinstatement purposes where the sum advised was insufficient to replace the Grade II property in a similar form to the original the plaintiff succeeded in a claim in tort for the amount of shortfall (*Beaumont* v. *Humberts* (1990)).

In another case of fire damage, tenants did not insure warehouse premises when entering into a lease on the landlords' assurance that they were held covered under a block policy. Subsequently, the landlords allowed the policy to lapse as a result of which the tenants were unable to recover monies when the building was destroyed by fire. It was held that there was no duty on the landlords to update previous statements although there has been some criticism of this harsh decision (*Argy Trading Development Co. Ltd* v. *Lapid Developments Ltd* (1977)).

Where rating valuers were engaged to advise on the rating assessment and had in fact secured a reduction they were nevertheless held negligent for not drawing to the attention of their client the implications of the rateable value in relation to a claim for compensation on leaving the premises. This decision seems to extend the principle of the duty of care to a worrying degree as the valuers' instructions appear to have been limited to the question of the assessment for rating (*McIntyre* v. *Herring Son & Daw* (1988)).

Other cases where property managers may have an interest in the course of their management activities, as opposed to a direct responsibility, are of some interest.

The negligence of a solicitor in failing to advise as to the existence of a right of way resulted in an award of the costs incurred in purchasing and operating the property. But a further claim for damages for anguish and vexation was disallowed as the contract was for carrying on a commercial activity with a view to profit (*Hayes* v. *Dodd* (1988)).

Interference with the normal enjoyment of their occupation of flats was caused to tenants in a block of flats by the construction of penthouses.

They were held to be entitled to loss of rental income from flats (*Mira* v. *Aylmer Square Investment Ltd* (1990)).

The ground floor of some commercial premises had a restrictive planning condition, rendering it unsuitable for the purposes of the client who acquired the lease. Their endeavours to assign the lease were unsuccessful and it was eventually surrendered to the landlord. The solicitors conceded negligence but contested the amount due to the tenants and also their liability for the costs of the tenants' surveyors but failed on both counts (*G P & P Ltd* v. *Bulcraig & Davies* (1988)).

Where a road contractor wrongfully dumped a large amount of spoil the question was as to the basis on which the damages should be assessed: was it to be based on the cost of removal or the depreciation in the value of the land? The former was the basis of damages awarded (*Minscombe Properties* v. *Sir Alfred McAlpine & Son Ltd* (1986)).

Plaintiffs who relied on accounts prepared for the company they wished to take over sued the accountants in tort when it became apparent that the accounts were wrong in a number of important respects. The accountants pleaded that they did not know that the accounts were being used for that purpose but it was held that they ought to have realised that they might be relied on, establishing a duty of care. However, the accounts had not caused the plaintiffs to do something which they would otherwise not have done and were not therefore the source of any loss; no damages were awarded therefore (*JEB Fasteners Ltd* v. *Marks Bloom & Co.* (1983)).

The question of the time at which damages should be assessed was important in a case involving damage by piling on adjoining land, when costs had risen considerably by the time the matter came to trial. It was held that the later date was the appropriate time for assessment (*Dodd Properties (Kent) Ltd* v. *Canterbury District Council* (1980)).

1.5.5 Limitations

The terms of many contracts seek to restrict the liability of the service provider by limiting any warranties given, particularly in respect of economic loss which is otherwise excluded from a claim for damages. Consequential damage is also excluded to avoid being responsible not only for the direct loss but for those losses of the particular claimant.

It is not possible to sue an arbitrator for the reason that he occupies a quasi-judicial role. Professional advisers do incur liability as do independent valuers. Some of the circumstances in which parties have been sued follow as illustrations of the extent of the liability incurred.

Where a diamond was displayed for sale in an auction room and was stolen, it was held that the respondents had not taken reasonable care. They were, however, held to be protected by an exclusion clause which provided that goods were accepted for sale at owners' risk (*Spriggs* v. *Sotheby Parke Bernet & Co. Ltd* (1984)).

Where the negligent spread of fire caused damage to premises not in use and likely to be demolished sooner or later the claim for damages failed. It was not a case for reinstatement and damages are intended to be compensatory. A plaintiff is not entitled to make a profit (*C R Taylor Ltd* v. *Hepworths Ltd* (1976)).

On the other hand, where negligent contractors caused the loss of a building by fire and the plaintiffs acted reasonably to mitigate their loss, it was held that the damages could not be reduced on the grounds of betterment (*Harbutts Ltd* v. *Wayne Tank and Pump Co. Ltd* (1970)).

Apart from exclusions agreed between the parties there is increasing statutory intervention in the contractual arrangements.

It was held in *County Personnel (Employment Agency) Ltd* v. *Alan R. Pulver & Co.* (1986)) that solicitors should warn their clients of the possible effect of an unusual clause that may involve risks to the clients. A firm engaged to advise on and negotiate rent reviews settled at rents below market levels and were held negligent in not obtaining sufficient comparable evidence on which to base their advice.

In *Radjev v. Becketts* (1989), the tenants appointed a surveyor but received no information from him about the proposals for rent on review; he did not submit any representations to the independent surveyor nor did he warn the client of the possibility that the rent might be fixed at a figure above that originally proposed by the lessors. The surveyor was held to be negligent. It was emphasised by the High Court judge in *Corfield* v. *Bosher & Co.* (1992) that solicitors should warn their clients of taking up the arbitrator's award and if necessary appealing against it within the statutory period on pain of being found negligent if they did not. Where independent valuers acting as experts under a rent review clause included in their valuation an area of 314 square metres of space which the lease had permitted to be removed to facilitate storage, it was held that the valuers were correct in adding in the area which was available, even though the particular tenants chose not to use it (*Hudson, A. Ppty Ltd* v. *Legal & General Life of Australia Ltd* (1986)). The tenant sued the independent valuer appointed to determine the rent on review in *Zubaida* v. *Hargreaves* (1993), alleging that the valuer had rejected one comparable and used other comparables of similar shop units. The case against the valuer was dismissed, the courts holding that he had carried out his duties competently.

1.5.6 Relevant legislation

Of particular interest to the property manager are the following Acts:

Health and Safety at Work Act 1974

It was held that common parts of a block of flats were a place of work (*Westminster City Council* v. *Select Management* (1984)) and therefore subject to the duty of the employer to ensure that the premises were safe and without risk to health. A claim of negligence could therefore arise out of the legislation.

Unfair Contract Terms Act 1977

Where any clause attempts to restrict liability it is subject to the statutory test of reasonableness in section 2. There is a need for clarity in the acknowledgement of instructions, the limitations subject to which they are accepted and in the report. Where sued in tort by a third party, the defendant may be able to rely on a clause excluding responsibility to any third party. Such an exclusion clause would rule out any special relationship.

A new law implementing EC Directive 93/13 and coming into force on 1 January 1995 makes further progress in the area of unfair terms in consumer contracts. Any contracts entered into from that date must be in plain, intelligible language and unfair terms will be unenforceable against the consumer. A contractual term will be unfair if it has not been negotiated individually and it causes a significant imbalance in the rights and duties under the contract to the detriment of the consumer. A test of fairness will take account of the respective bargaining positions of the parties and whether an inducement was offered to enter into a particular term.

Supply of Goods and Services Act 1982

The Act refers to any contract under which a person agrees to carry out a service and will therefore include any property management arrangements.

The exercise of reasonable care and skill is required (section 13). Where no time is fixed for performance, then a reasonable time is implied by section 14. It is also implied that the person receiving the service will pay a reasonable charge. Section 16 provides that implied duties may be varied by agreement between the parties but subject to the overriding test of reasonableness under the Unfair Contract Terms Act of 1977.

1.6 THE FUTURE

Much of the property in use is old and unsuitable for the purpose. With increasing demands and standards, substantial numbers of buildings will be vacated or fall into disuse. Many buildings occupy restricted or otherwise unsuitable sites, having little residual value. The inevitable consequence will be a major structural movement accompanied by a demand for more suitable and efficient buildings.

1.6.1 Need for improved market intelligence

The demand for skilled, capable, imaginative and well-resourced managers is likely to increase and flourish. Even allowing for real differences between the property market and the market in stocks and shares, there is some way to go before the sophistication of the investment analyst is matched in all sections of the property field. Some advance in technique interpretation and application will be demanded by investors cognisant of the effect of good management on performance and therefore on value.

1.6.2 Expectations of occupiers

Industry and commerce have experienced very severe trading conditions and will be reluctant to take undue risks in new ventures. Greater flexibility in the ways in which property could be occupied and where the ongoing penalties for failure were less draconian and long lasting would be likely to encourage more businesses to take on premises for expansion of existing trading or to develop other activities. In this quest, shorter initial terms, inclusive rents, options to surrender and the absence of a continuing liability on the part of the tenant or personal guarantors would all contribute to the tenant's decision.

1.6.3 Simplification for investors

The property market has recently undergone the most severe structural change in its history, the full effects of which are yet to be experienced.

Many buildings are unsuitable for any use and not capable of being brought up to current standards, whilst redevelopment is uneconomic. The concentration of activity on fewer sites, the advanced state of automation in many industries and in the distribution and service sectors and the consequent employment of fewer workers all point to the demand for space being of a different order. At first glance these movements would seem to restrict the market even more than at present. This need not be so. Although discussions on unitisation and securitisation intended to widen market appeal were inconclusive, they served a useful purpose in exposing the problem to public debate: securitisation has refinanced a small number of schemes in the United Kingdom and is widely and successfully used in the United States of America where the introduction of a financial guarantor with a high credit rating absorbs much of the risk in exchange for a premium and thereby enhances the standing of the investment.

1.6.4   Other forms of investment in property

There are indications that property derivatives would be welcomed by the market. A recent issue by Barclays Bank of £150m of variable short-term property index certificates (PIC) provides both an annual yield over the life of the investment and a performance bonus on maturity measured by reference to a property index.

FURTHER READING

Armstrong, M. (1990) *Management Processes and Functions* 1st edn, Institute of Personnel Management.
Jackson, R.M. and Powell, J.L. (1987) *Professional Negligence*, 2nd edn, Sweet & Maxwell.

Policy formation and implementation

2

Unless there is an agreed policy there is a danger that much management activity will have little purpose and the investments will underachieve. This chapter discusses the objectives of estate policy and the support needed for its development and implementation. It considers the prerequisites for policy formation, outlines the approaches to decision analysis and touches on portfolio theory.

It also acknowledges the difficulties in strategic planning which are inherently greater with direct investment in real estate than with other investment media.

2.1 INVESTMENT IN PROPERTY

The objectives of property management are to maximise the net return whilst preserving and enhancing the capital value and future of the subject matter.

The return may be variously measured in terms of income and income growth or capital appreciation or a combination of the two, in transformation of an asset by development to achieve a different and enhanced asset or to achieve any other stated objective which would normally have a direct financial outcome but not necessarily so. The process has many similarities to any other management task, in general requiring the definition and evolution of policy and its implementation.

2.1.1 Optimisation levels

Investment optimisation is effected at two levels, the routine and the strategic. The routine consists of regular collection of rents and charges, inspections, ensuring a satisfactory level of maintenance (whether at the expense of the occupier or owner), maintaining insurance cover at a sufficient level, ensuring that the lease covenants are complied with, initiating rent reviews and reletting premises that become vacant. The strategic embraces posing the proper questions and seeking the answers to ensure that long-term decision making is informed by good quality, reliable information. There is no clear division between routine and strategic; reletting a vacant property may at one and the same time be a routine matter directed towards maintaining the income flow and a strategic one given that the vacancy offers an opportunity to consider the various options at a time when freedom of action is at its greatest.

2.1.2   The range of tasks

The overall task requires the exercise of common management skills adapted to the property scene. In a well-regulated company, stewardship would operate at several levels. The daily repetitive tasks such as accounting and compliance matters would be regulated by staff with a good technical background. Competent property managers would bring to their task in addition a detailed knowledge of all aspects of the property market, an understanding of value and the valuation process, a grounding in land law and an ability to anticipate and exploit the potential of property at each stage in its life cycle. The professional investor is likely to have more than one building or property and the final requirement of the property manager is an ability so to arrange the collection or portfolio of property as to maximise the return, however defined, in a way that also acknowledges the need to contain risk to an acceptable level.

2.1.3   Service to the individual client

The precise objectives and their relative importance will be a matter for discussion with the individual client and will vary from client to client. Before appointment, the property manager has probably indicated the main thrust of their initial views on the portfolio which he or she will now expand in co-operation with the client. The overall objectives of maximum return and aversion to risk common to most investors do not necessarily govern all parts of a portfolio at all times. A large, old-established and well-run fund with low gearing may well be prepared to invest a modest percentage of total assets – albeit a substantial sum – in a more risky proposition if recommended by its advisor: for example a development proposal with a low initial return where the longer-term situation looks attractive and secure or, possibly, where such an investment could protect or enhance other parts of the portfolio without itself necessarily providing anything near the required overall yield. Similarly, a lower return on a particular investment may be acceptable where substantial tax advantages, long-term enhancement or other benefits are envisaged that compensate for the lack of an adequate current income.

The task of property management is not repetitive or predictable in the way in which an item emerging from the production line in a factory is repetitive; there are recurrent tasks such as rent collection, maintenance inspections and the like that are routine matters and benefit from being organised in that way, though even then it requires alert and responsible management to identify particular problems before they become unmanageable. Property management should respect the fact that each property making up the holding is unique and undergoes substantial change over time. Maximisation of potential calls for further abilities of resourcefulness and imagination to ensure that each stage in the life of a building and the land on which it stands is recognised and exploited.

2.1.4 Extent of property manager's authority

There has been a limited trend in recent years for property managers to be appointed on a discretionary basis whereby they are given the power and responsibility to manage the asset without reference to the client. But in the majority of cases, property managers will have only limited authority to make changes having major financial or policy implications, particularly where the management is of part of a fund which has substantial assets and is supervised by in-house professional management: in such cases their priority must be to obtain clear and firm instructions based on sound, well-reasoned advice.

2.2 DEVELOPING AN INVESTMENT STRATEGY

The majority of investors will have a mature understanding of financial markets, although those without detailed experience of property investment may need to be reminded of the crucial differences between holdings of stocks and shares and direct investment in property. The first requirement is to recognise that frequent change is impracticable and uneconomic, underlining the importance of careful initial selection in relation to a strategic plan. There is a need to buy and sell to maintain and enhance the quality of any portfolio; the more precisely and strictly are the attributes of a desired portfolio defined, the shorter will be the holding period. It should be accepted that the asset is illiquid and in general indivisible and there is the associated problem of reinvestment of any part of the realised amount not otherwise earmarked for use on the fund's existing assets. Furthermore, the investor needs the willingness and ability to react to unscheduled opportunities that may occur at the most unexpected and inconvenient times.

Each property has a unique location and other legal and physical aspects distinguishing it from other property, even where superficially similar.

Market research is not only more difficult to conduct but the outcomes are more difficult to relate to other situations. Advances have been made in recent years to the point where there are now eight or nine major monitoring programmes – albeit with different bases and emphases – reporting performance measures including capital growth, rental change and yield movement based on market transactions and valuation information.

2.2.1 Range of information needed

The manager will therefore need to research the particular client's needs before he or she can propose an appropriate policy. The precise information will depend to a large extent on the present size and nature of the portfolio and the availability of further funds to achieve any aspirations for growth. A vast amount of information is required before the exercise can be undertaken; the following list suggests some of the more usual information although its importance will vary from one fund to another:

(a) The collection of properties (factual)

- size of portfolio
- type of individual property
- geographical spread

(b) Physical

- age of buildings
- quality of buildings
- type of buildings
- specialised buildings or uses

(c) Financial details

- capital value
- rental income
- covenants of occupiers
- capital available now and in the future

(d) Legal details

- nature of legal interests
- authorised planning use class
- lengths of leases
- terms of leases

(e) Opportunities and prospects

- redevelopment/upgrading
- possible disposals
- possible rearrangement of leases

(f) Deficiencies requiring attention

- underachievement
- unacceptable level of risk

(g) Other considerations

In formulating a policy for discussion, approval and adoption the property manager will need to consider and advise on a number of issues:

- state of current holding – whether sound, weak, deficient, etc.
- current yield – whether appropriate to risk, whether likely to be sustained; opportunities for improvement
- appropriate size of individual investments – lumpiness
- mix/balance/rebalance
- disposals
- acquisitions
- renegotiations
- enlargement of portfolio
- targets

2.2.2 Advice on policy

Where advice on policy is within the remit of the property manager he or she will develop an estate policy document having regard to the needs of the particular client. For example, a private trust may have no need for income for the next twenty years and may therefore undertake major investment on parts of the estate to achieve added value and income growth in the longer term whilst accepting that there will be little or no current surplus income. An insurance company committed to providing pension payments will wish to match its income to actuarial projections of its commitments in future years; part of its policy could therefore take account of reversionary investments showing a low return for some years to come. In such ways, the shortcomings of the existing portfolio would be highlighted and identify the nature of the holdings where emphasis should be given.

Selection is paramount; unless the right decisions are made at the investment stage, there is only a limited amount that can be achieved through strategic planning.

2.2.3 Adoption of policy

Once the strategy has been developed and costed, it is submitted to the client for discussion, possible amendment and eventual confirmation. Adoption of such a document as a policy would include a recognition of the level of expenditure and change required to achieve it; the policy must then be kept under constant review and, if necessary, adapted to internal and external influences likely to affect it. It should be regarded as a blueprint that will only be effective if it is maintained in this way and accepted as a broad directional indicator rather than a precise route. In particular, it should not inhibit consideration of unexpected opportunities. Where the fund is a general one, the question of asset allocation will have to be addressed. A case has been made for the allocation of up to 20% of the available funds of a large mixed fund to be placed in property, but very few fund managers would be prepared to commit funds to that extent. The average allocation in pension funds was 18% in 1981 and has now fallen to 6%: with increasing demands on pension funds and the need for greater liquidity as one of the outcomes of the Goode Report, it is unlikely that property will play a significantly greater role in the investment policy of funds in the foreseeable future.

2.2.4 Contingent liabilities

Contingent liabilities may exist as the result of terms imposed by a head lease or conveyance or where expenditure is likely to be necessary to comply with statutory provisions. Where the outlay is likely to be substantial and certain to arise at some time in the future, owners will wish to budget for its impact. They may either make provision by earmarking internal funds, or consider setting up a sinking or reserve fund sourced externally. Should the opportunity arise to remove a liability (by purchase, merger, negotiation, exchange or otherwise) it deserves serious consideration even where no immediate financial benefit is derived.

Where a liability is based on the existence of a legal limitation, such as a restrictive covenant entered into many years ago, probably by a predecessor in title, it may be possible to insure against the prospect of a third party claiming the benefit of and seeking to enforce the covenant.

2.2.5 Life cycle

Consideration of the life cycle of land will recognise depreciation of the built part of the asset and the effect of change on the contribution of each component of the portfolio.

Although property may be viewed as long-term, even permanent, investment because of its reassuring physical presence, it is this feature that adds a further dimension to the ownership of and investment in property. Even if the building represents a good example of current standards of comfort, design, construction and equipment when built, it will then become prone to obsolescence. Physical obsolescence leads to depreciation and lower values but can be countered by a programme of maintenance investment including replacement of equipment and facilities that have a shorter life than the building itself. Developments in usage or layout may lead to functional obsolescence which may be difficult to counteract whilst the ageing of the building or a shift in preferences for other areas (of the town, of the country) for the activity carried on may lead to economic obsolescence. These are all risks inherent in the original investment decision which will recognise that, apart from any accelerated obsolescence, each property has a natural life cycle. The general model of a typical life cycle is suggested in Figure 2.1 although real life situations are far more complex and variable.

2.2.6 The economic cycle in property

An important feature of any market activity is its cyclical component. Briefly this is the period over which any economic good completes a full cycle of price movement from one major low to the next (from which most economists usually exclude minor trends lasting less than one year).

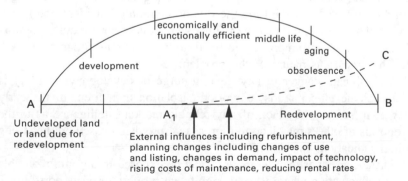

Figure 2.1 Typical life cycle for property.

The property market suffers from the phenomenon at least as much as stocks, shares, commodities or currencies. Some commentators claim that the property cycle is contracyclical, that is that it occurs out of step with movements in other investment media, using this argument to support the notion that a portfolio should contain a proportion of direct property investments in order to achieve a better balance. Whilst not dismissing support for inclusion of property the claim may be too strong in that property feeds off other economic activities and is affected by the strengths and weaknesses of other sectors. Much of the accentuation may be attributed to one of the distinguishing features of property investment: the existence of the institutional lease with its long-term commitment by the tenant to upward-only reviews, a defensive feature in any downturn. Such an admittedly anti-market structure limits risk and shields the investor from some of the effects of a recession in the short term but where that recession is long and deep and particularly where it has structural elements, the rigidity of the arrangement itself may well precipitate bankruptcy or liquidation of the tenant.

The development process is a long one and adds further uncertainty. Any proposal reaches a point quite early in its life where either it cannot be aborted or it is prohibitively expensive to do so, as it would involve frustration of contracts and actual or notional holding costs without any compensating income. The commitment to development following identification of an opportunity may be followed by a marked downturn in the market generally or an easing of rental values and yields arising directly from the overall development activity. In a situation where the cost of an individual development project may run into millions of pounds and where total investment is measured in tens and hundreds of millions, many development decisions are nevertheless based on inadequate information and flawed forecasting in determining the risk of producing a building for which at the time there is no committed or identified tenant and of which the rental income is uncertain. It is of course quite possible that the market will improve and result in a higher return than estimated, but the precarious nature of the problem remains.

A significant danger in any general surge of development activity is that each developer tends to work in isolation and often, in the early stages, in secrecy; the lack of any reliable market intelligence as to the proposals of others may lead to a gross overprovision based on the strong initial indications of demand.

Further, an investment company may be exposed to any unfavourable market fluctuation where it has a significant proportion of its portfolio as land for development, contemplated or in progress. The established investment institutions are strong enough financially to carry the funding and other costs of a development that is not immediately successful but will seek to avoid undue risk by committing only a small proportion of total funds in this way. Institutions are not usually leaders in identifying development opportunities and often approach such a prospect as a joint venture, thereby further limiting their exposure.

Additional uncertainty in property development and holding is brought about by changes threatening the market. First, there is the pressure for different lease structures tending towards shorter terms and a more equitable position between the parties. Increasingly, tenants wish to be relieved of the direct responsibility for repairs and maintenance whilst being prepared to discharge that liability through a higher inclusive rent. The valuable action of distress has been criticised and may not survive as a special and very effective form of debt collection whilst moves to protect tenants who have assigned their interests but remain liable in the event of the assignee's default have been shelved for the time being but are likely to be raised again. Secondly, there is speculation about the eventual location of major shopping areas – regional developments with a wide range of shopping provision with good access and extensive parking – or town centre with the constraints of the original or modified street plan, the intrusion of vehicles for delivery and the difficulty and cost of parking in what shoppers will accept as a convenient position. Distribution points gain much by adjoining a motorway or major road with access to the main system; offices were thought to be at risk from the development of information technology but so far this has not proved to be the case. Some firms have looked more closely at usage resulting in a dramatic reduction in space allocated to salespersons and others who spend much of their time out of the office visiting clients and travelling.

Thirdly, there are new requirements and standards much discussed but less often achieved. The pointers to quality exist but are often ignored by developers in pursuit of the highest immediate return. Occupiers often spend considerable sums in fitting out but at that time are constrained by the basic structure. Important issues in the design and operation of a building are floorspace, flexibility, adaptability and costs in use. At the same time, there is no point in overprovision especially as it invariably results in additional costs. The question of the optimum life appropriate to a building needs consideration, particularly in the scope for reducing initial cost. External factors include increasing concern for environmental issues, in particular continued use of the motor vehicle in built-up areas and the effect of stricter controls on land contamination.

Some companies large enough to contemplate property investments nevertheless exclude them from their property portfolio. Where property forms part of a portfolio, the average asset allocation is currently around 6%, even though there is broad agreement that property should reasonably represent up to 15% of the total allocation.

2.3 INVESTMENT CRITERIA

The investor wishing to place some funds in property will take account of the need to manage that investment and the greater effort needed to understand the market. This is inevitably so since each property has a unique location and other legal and physical aspects distinguishing it from other property, even an apparently similar one close by. Research is not only more difficult to carry out (where there are relatively few transactions at fairly high and indivisible unit prices) but the outcomes are more difficult to apply to other situations. Improvements have occurred in recent years to the point that there are now some seven or eight monitoring programmes reporting various performance measures including capital growth, rental change and yield movement based on market transactions and valuation information, the largest of which is Investment Property Databank (IPD).

However the point needs to be made that much of the information gathering is weighted towards the larger and better investments and is therefore not entirely representative. The market in stocks and shares is immediate and transparent, at least in terms of price, whilst the effect of economic events is reflected in the price on a daily basis and can therefore be charted accurately. The property market remains fragmented, incomplete and affected to some extent by interpretation; consequently, any conclusions should be viewed with a certain amount of caution.

Contingent liabilities may exist as the result of terms imposed by a head lease or conveyance or where the expenditure is necessary to comply with statutory provisions.

Where the outlay is likely to be substantial and certain to arise at some time in the future, the owner should consider building up an adequate reserve or sinking fund sourced externally or, more likely in the case of a large group, provided from internal resources. Where there is a liability based on the existence of a legal limitation, for example a restrictive covenant entered into many years ago, probably by a predecessor in title, it may be possible to insure against the unlikely event of a third party claiming the benefit of and enforcing the covenant.

Should the opportunity arise to remove a liability (by purchase, merger negotiation or exchange) it should receive serious consideration even though no immediate financial benefit is achieved.

2.3.1 Investment in existing property

Although the aim will be the construction of a perfect portfolio – 'perfect' being the definition of the size, location and type of holding specified by the investor or their advisor – such a state is likely to be an objective rather than a reality. In short, it is much more difficult, expensive, time consuming and risky to achieve and maintain the optimum holding where the investments comprise property interests.

2.3.2 Renewal by refurbishment or redevelopment

A particular aspect of property and the need to ensure a proper return on capital is the equation of demand and supply in the particular circumstances of the property. At some point as rents increase the developer or other supplier of accommodation will deem it appropriate to enter the market to supply more space at the prevailing price (rent).

Present economic conditions and the state of business confidence are not conducive to schemes of comprehensive development. Whilst the level of building costs has risen in recent years at less than the rate of inflation, the uncertainty of trade and commerce and the difficulty of pre-letting have added to the risks attendant upon redevelopment. The alleged relationship between inflation and growth is no longer pursued as vigorously as once it was and it is no longer seen as inevitable, with many examples of reduced rental levels. Those properties purchased at an optimistically low initial yield are unlikely to achieve the rental growth implied by the price paid and will require all the expertise and ingenuity of their managers to avoid the worst embarrassments of any failure to achieve the growth implied by the price paid.

The long lead-in time between identifying a profitable opportunity and the completion and letting of a building can often be measured in many months or years during which the market is likely to change in significant ways. The further unique problem is that other developers may identify similar opportunities and react at about the same time; there is no reliable monitor of the demand–supply equation and, once land has been acquired, it becomes progressively more difficult and expensive to abort the development process. By the time the intentions of others are known and the effect on the market as a whole can be judged, decisions have probably been taken which already commit the developer to a particular course of action. A particular development may be promoted as satisfying particular needs but there are almost always alternatives for both providers and users (including doing nothing). Nevertheless the market becomes active, the general euphoria extends from developers to banks and other financial institutions all determined not to be left behind until, if the anticipated demand has been overestimated or overprovided for, the anticipated rental level may not be achieved and there is a significant amount of empty floor space incurring financing cost, maintenance and insuring responsibilities, management charges and rate liabilities.

2.4 INFORMATION BASE

A prerequisite of any approach to setting out proposals and determining a strategy for a portfolio is the availability of information. There must be a reliable database sufficient to enable the units within the portfolio to be assessed for suitability, initially without inspection.

The information collected should be sorted and stored methodically. As indicated above, there is a considerable body of material of possible use, best collected systematically rather than left until a specific requirement arises. The tasks of collecting, sorting, storing and retrieving data has been much assisted by the advent of low-cost computer hardware and powerful but relatively simple software. Photographic quality records, plans and other details may be stored graphically and should contribute to greater efficiency and lower costs. The topic is treated in greater detail in Chapter 4.

2.5 DECISION ANALYSIS

Decision analysis may be described as the systematic approach to a problem or a situation where more than one solution is available, each of whose outcomes is uncertain to a greater or lesser degree. The process of decision making has become much more sophisticated in recent years, investment desirability being heavily dependent on the two key criteria, expected returns and degree of risk. Decision making should be structured and positive. The main concerns of most investors are the expected return and the level of risk associated therewith. One tends to influence the other but as most investors are risk averse, the return is likely to reflect the determination of the investor to avoid any undue risk.

The simple act of investment in property itself involves a risk but there are ways in which it can be contained or reduced.

2.5.1 Judgement

A particular course of action is not necessarily the only course of action available: many decisions involve a choice of action from a number of viable alternatives each of which might be acceptable in isolation. It is possible to refine the decision and reduce the risk inherent in its implementation by ensuring that the information used is relevant and it will be seen that its selection is in itself a judgement and therefore part of the process of reaching a decision. In the early stages, great efforts should be made to avoid bias or prejudice in selecting what is relevant and what is not.

It should be emphasised that a judgement is no more than an informed opinion based on the collected material and that any change in that material or its interpretation through experience or otherwise or the emergence of new material requires at least a reassessment of the decision and points to the need for vigilance and flexibility in ensuring that directions derived from judgements remain valid. The process of judgement is not well understood despite much study. It is axiomatic that experience and education contribute to the quality of judgement but it is more difficult to explain the inherent ability in some – variously described as inspiration or flair – to reach a sound conclusion apparently without a formal deconstruction of the problem. Each approach has something to contribute and may be the basis of the 'brainstorming' approach where the rapid and uninhibited flow of ideas each encapsulated in one or two key words stimulates cross fertilisation of views, often resulting in a wider exposure of possible actions. In different people the reception, selection and interpretation of words, figures or visual images tends to vary, which will be an asset in the process. Care must be taken to keep preconceptions under control and to avoid the tendency to obtain the anticipated result. At the end of such a session there remains the need for one or more of the group to go through the process in a more considered way to develop the embryonic solution proposed.

The next stage is to distil the information to determine its validity and the weighting each item should be given. Some information will be factual, some will be partly supported and some will be based on no more than inferences or assumptions, incapable of clarification at that stage. The objective is to provide reliable advice based on a thorough assessment of all the available information.

At this point, the adviser or decision maker has a great number of contributions on which to base his or her recommendation; it is perfectly acceptable to elect to take no action as a result of the information disclosed.

2.5.2   Approaches to analysis

Decisions are best founded on well-reasoned argument supported by the rigour of mathematical demonstrations of the likely effect. Many decisions still rely on judgement only, an approach sometimes described as qualitative analysis, although given the absence of any formal analysis, this is a somewhat misleading term.

The following brief descriptions are intended to do no more than provide an awareness of some of the techniques available. Standard texts on financial decision making provide detailed mathematical formulae and explanations for those wishing to pursue the topic, whilst there are many financial and statistical computer packages available.

(a)   The payback method

The name aptly describes the intention to rate the capital invested in a particular transaction according to the speed with which the incremental cash flows replace the original cost. The lower the numerical result, the more favoured the proposal. The method is intended primarily for short timescales and has little serious application to property investment decisions since it disregards both the time value of money and the residual value of the investment and avoids the question of risk. In essence it describes the quick year's purchase calculation made instinctively by investors in the qualitative approach described above, acknowledging that it is no more than a first impression requiring a considerable degree of refinement before it has any credibility in informing a decision.

(b)   The return on capital employed approach

The return on capital employed (ROCE) approach is also known as the accounting rate of return and is, as its latter name implies, a measure of performance used by the accounting profession. It may be expressed either by calculating the ratio of average annual profit to the average capital employed over the life of the project or the ratio of average annual profit to the initial outlay (which by definition ignores any increased working capital required to support the operation over its lifetime). The limitations are similar to those of the payback method whilst this approach cannot be modified to reflect the time value of money although it is possible to build in residual benefits or costs. The main disadvantage with both approaches is the preoccupation with profit at the expense of the underlying capital value of the asset. Whatever usefulness it may have appears to be limited to short-term projects but neither would cope with the complexities of assigning a value to a short run leasehold interest.

(c) The discounted cash flow approach

There are two generally accepted appraisal techniques, both using discounted cash flow methods, thus overcoming two objections to the payback and ROCE approaches, use of the time value of money and reference to a rate of return. They are the net present value (NPV) and the internal rate of return (IRR).

(i) Net present value technique

The initial and estimated future outflows are compared with the anticipated future inflows including any terminal inflow, all discounted at a predetermined rate of interest. Then, if the result shows a positive or zero net value, the investment may be regarded as acceptable if it meets the stipulated criteria. Determination of the interest rate is associated with the return available from alternative investments and supposes that any higher initial cash investment would make the project uncompetitive. (It will be noted that the interest rate is arrived at by reflecting the risk inherent in the project.)

(ii) The internal rate of return approach

The cash outflows and inflows are compared as previously except that this model indicates the discount rate which produces a zero NPV. The rate deduced may then be compared with the expectations of the company or fund in determining whether the investment is one which it wishes to pursue.

The result in either of the last two cases may be obtained from mathematical calculations or graphical plotting and both may be obtained by use of a computer program.

It should be noted that under certain conditions an investigation may show an investment to have more than one internal rate of return or none at all. This is particularly likely to occur where there are constant changes between cash inflows and outflows resulting in changes of sign with which the polynomial theorem on which it is based is unable to cope. Resolution of the problem may involve isolating some of the cash flows to eliminate the changes in sign or substituting the NPV technique, using the desired rate of return.

2.5.3 Multiple criteria decision making (MCDM)

Many of the existing aids to decision making are designed to cater for very limited criteria, typically being limited to risk and return. More recently, the need to take account of further considerations has led to the evolution of multiple criteria decision making or decision aid (MCDA). The concept is a combination of mathematical calculations and information related to the decision maker's preferences in the general case. It is a development of earlier processes rather than a new process and is facilitated by developments in the field of expert systems and computer software generally.

2.5.4 Expert systems

Knowledge-based or expert systems are computer programs devised to replicate the reasoning process or logic employed by human beings. Early work in this field derived from research into artificial intelligence and concentrated on rule-based knowledge. The design to emulate human expertise derived from a knowledge base provided by the user and capable of dealing not only with information and facts but with judgement also. An important feature is the ability to grow by storing the responses to questions posed by the program in the process of resolving a problem which are then incorporated in the knowledge base.

There are two immediate needs: the design of the program for the particular application and the co-operation of and acceptance by the user. The design of a particular program is facilitated by purchase of a commercially developed expert system shell, which gives shape and form.

A simple example concerning a business tenancy indicates the nature of the system which uses expressions such as those in Figure 2.2.

The program is then able to interrogate the user to obtain further information on which to base the advice. In this example, it could be programmed to venture into the area of judgement, expressing an opinion as to the advisability or otherwise of serving notice, taking account of the current rent level and the availability to the tenant of alternative accommodation. Similar use could be made of a system in relation to decision

IF	the tenancy is subject to the Landlord and Tenant Act 1954
AND	the end of the lease is within one year from now
AND	the tenant has not served a section 26 notice
THEN	the landlord should consider serving a section 25 notice.

Figure 2.2 Simple example of an algorithm in an expert system.

analysis or investment performance, including any mathematical or financial calculations but also offering advice.

The initial reaction of users is likely to be scepticism, possibly hostility or suspicion. Should this be the case, they should be encouraged to reach a conclusion by their normal route and then compare this with the outcome suggested by the program. As the answers either confirm the user's view or ask valid questions which had not occurred to them and which change their opinion, they will reach a judgement about the value of the system and possibly engage in further programming to enhance its ability.

2.6 ASPECTS OF PORTFOLIO THEORY

The principle of portfolio theory is that investments should be spread across a range of suitable holdings, thus achieving a reduction in risk. It is assumed that investors act in a rational way and will accept the highest return compatible with the level of risk they are prepared to accept. Where the same return is available from two potential investments, the investment with the lower risk will be preferred. In all cases, they will look for the best combination to maximise return and minimise risk. This general principle has long been recognised and is encompassed in the advice not to put all one's eggs in one basket.

Portfolios which minimise risk in relation to return are termed 'efficient' and when the holdings proposed for or held in the portfolio are plotted on a graph may be used to indicate an efficient frontier. The expected rate of return is the weighted average of the components, the proportions invested in the various securities being used to determine the weighting. The degree of risk attaching to the return on each component is then combined to measure the correlation coefficients across the portfolio. The larger and the more carefully selected the holding, the greater the negative coefficient, which is another way of saying that the risk associated with one of the components is offset by a contraindication with another; the range between perfect positive correlation and perfect negative correlation is +1 to −1. In general, the higher the positive correlation of individual components, the greater the variability of return, leading to a higher standard deviation.

With proper selection the overall risk is reduced when investments are held together in a portfolio; in other words, the portfolio risk is less than the weighted average risk of the individual components because of the compensation between investments where the risk does not change at the same time and in the same direction. In considering risk reduction in a property portfolio, the adviser would seek negative correlation by selecting investments across the spectrum, for example choosing different sectors and different geographical areas.

There are difficulties in applying the theory to any investment, but particularly to property investments which contain the additional element of physicality. Further, much of the return is not actual but anticipated and misjudgement of the level of a rent review or of how long a vacant property will take to let may have a serious impact on the overall return of the component.

Modern portfolio theory is a description reserved for the work attributed to Markowitz who introduced a mathematical rigour to the development of this branch of financial analysis in his writings throughout the 1950s.

2.6.1   The capital asset pricing model (CAPM)

A further development of portfolio theory is the capital asset pricing model, the main endeavour being to find a simpler model and therefore one likely to be adopted more widely. The model measures systematic or non-diversifiable risk, being risk which is related to the market and therefore cannot be diversified within a portfolio. The model measures the sensitivity of an investment to market fluctuations, increasing volatility being indicated by increasing results of the measurement.

Although some research has taken place in relating this model to property portfolios, there is a deal of scepticism about its possible usefulness in this regard, largely based on the practical difficulties of measurement and the differences between property and other investment media.

2.7 MEASUREMENT OF PROGRESS

Direct property investments are not appropriate for short-term investors. The costs of acquisition and disposal are higher than for most other investment media and frequent disposals would simply reduce the net yield. The illiquidity of property investments makes opportunist sales less likely than in shares where a sudden and substantial increase in price can be taken as a profit quickly and with certainty and equally quickly reinvested. The client will wish to know the result of the management undertaken and especially if it has been successful. The next chapter is devoted to the measurement of performance.

FURTHER READING

Drucker, P.F. (1994) *The Practice of Management*, Butterworth–Heinemann.
Henry, J. (ed) (1991) *Creative Management*, Sage Publications.
Northcott, D. (1992) *Capital Investment Decision-Making*, Academic Press.
Rees, W.D. (1988) *The Skills of Management*, 2nd edn, Routledge.

Property performance: evaluation and projection

3

Information on the performance of stocks and shares is available on a daily basis, grounded on actual transactions. Each unit is homogenous, liquid, accessible and is traded within a relatively perfect market. By contrast, property is unique, of varying lot size, indivisible and illiquid. The difficulties in constructing suitable and acceptable measures of performance are accordingly greater.

3.1 INTRODUCTION

Investors in the main financial securities – government stocks and bonds and equity shares – have long had access to detailed technical information of the price and volume of sales on a daily basis coupled with comment at various levels by financial journalists and retailed advisers.

3.1.1 Recent background to property investment

By contrast, little information was available regarding property performance until the early 1970s. As life assurance companies and pension funds then began to take an interest in direct property investment, the need for information assumed a greater degree of importance to enable those institutions to fulfil their duty of due diligence towards their members, the eventual beneficiaries. The requirement was heightened by the market crash of 1974 brought about by high inflation and interest rates, a reduction in consumer demand leading to a reduction in demand for new premises and the imposition of development gains tax on first lettings, creating uncertainty against a fragile background. The election of a labour government and its announcement of a more comprehensive development land tax completed the loss of confidence.

A number of property companies failed and many more would have done so had not the government prevailed upon the Bank of England to organise a 'lifeboat' forcing the joint stock banks to provide financial support for the secondary banking system. The ensuing period saw heavy losses, company failures and redistribution of assets, although the full extent of the problems was not always visible.

A new generation of developers failed to see the warning signals of the market crash in 1987 and although it claimed many casualties, this time the major banks were the principal lenders and were able to survive although they found it necessary to write off substantial losses, accentuated by injudicious lending. It has been reported that Barclays Bank Ltd, the most active in lending on commercial property, made provision for some £2.5bn in bad and doubtful debts in 1992.

3.1.2 Operational property

It has been noted that some well-publicised takeovers of companies made companies look at their property assets in a different light.

There has been a sharp increase in the attention to performance in recent years by investors in their desire to achieve as high a return as possible and at least comparable to that obtained from other investment media. But the 1980s not only found a more demanding and searching private investment sector, but a recognition on the part of public corporations and the boards of companies owning operational land, that targets were also applicable in their operations, albeit applying different measures.

3.2 PUBLISHED MARKET INTELLIGENCE

A number of the national and larger regional firms in the profession had published descriptive market reports for many years. It became the practice of the *Estates Gazette* to publish a selection of practitioners' reviews of the previous year each January. None pretended to a scientific basis; nevertheless many useful comments could be culled from the accounts and a few took the opportunity to make predictions about the market in the forthcoming year. A number of the larger, mainly national, firms then started to publish market reports at regular intervals, indicating rental values and yield movements which together would give an indication of capital growth. Some of the early reports were little more than public relations exercises, the information on which they were based rarely being sufficiently defined or rigorously analysed to deduce anything but the most general and often obvious conclusions.

The reporting scene has matured and several companies support significant research activities. There are now some seven or eight main indices (ignoring several produced solely relating to the market in agricultural land, forestry and sporting rights). The best known are cited in Table 3.1. The largest and most widely known is that produced by the Investment Property Databank (IPD) sponsored by, but independent of, 15 leading companies of surveyors. The database includes information from 120 funds, reflecting the performance of over 23 000 properties with

Table 3.1 Market reports

Source	Type of Report
Healey & Baker	Market reports – rent and yield performance and change
Investors Chronicle/Hillier Parker	Rent indices, yields and market values
Investment Property Databank	Monthly and annual indices of return, capital and rental growth by sector and geographical region
Jones Lang Wootton	Wide range of performance information
ML/CIG	Total returns and range of performance measures
Richard Ellis	Total return, capital and rental growth (world-wide)
Weatherall Green and Smith	Quarterly and yearly indices by sectors

a total value of some £41bn. A subsidiary index charts the movements of 50 unitised property funds, reflecting the performance of over 1750 properties with a value in excess of £3bn. Much of the remaining research relates to specific aspects or sectors of the market: for example, Hillier Parker publish quarterly reports on property market values, secondary property, specialised property, rents (contained in an index) and average yields, also a European Property Bulletin. Healey and Baker publish PRIME, a quarterly review and statistical analysis embracing the main market sectors (retail, out-of-town-retail, office, industrial) and national rental values (all relating to 'prime' property, which is defined as '... located in the best position within a monitoring point, are built, designed and maintained to the highest standards then applicable in the market. They are of the most marketable size and design and are let to tenants of unquestionable covenant on modern lease terms').

Many publish occasional papers, some based on a considerable depth of research, on a wide range of issues of importance and interest to the property world.

3.3 THE DEMAND FOR INFORMATION

There has been a sharp increase in the attention to performance in recent years, first brought about by the scrutiny of investors and their desire to achieve a target yield as high as possible and at least comparable to other forms of investment with similar levels of risk. The 1980s found a more sophisticated and demanding private investment sector and also a recognition that targets were applicable also to operational land whether held in the private or public sector, albeit using different measures. Some well-publicised takeovers of companies, not in the course of growth or integration or for their intrinsic worth as businesses but for their value on break-up where the land and property assets – especially where the buildings were old and the nature of the surrounding area had changed – made companies look at their property assets in a different light. Public ownership also came under scrutiny, being attacked on two fronts: is the land necessary for the purpose for which the organisation exists and, if so, is it being used in its optimum form?

The data required for a comprehensive analysis of activity in the property market extends beyond that required for the analysis of the prices of stocks and shares and its interpretation is more complex. Rent, rental value, rental growth and equated yield or target rate of return enable the total return to be calculated.

3.3.1 Principal data required

A brief description of each contributor follows.

(a) Rent

The rent is the current net rental income which may be the same as or less than the current rental value or in some cases in excess of it.

(b) Rental value

The rental value is the net rent likely to be achieved if available for letting on the open market.

(c) Rent reviews and lease renewal information

The number of years before the rent is next reviewed and the interval between reviews are both pieces of information required in performance measurement.

(d) Estimated yield

The expected growth explicit yield is the target rate of return which includes an allowance for risk accepting gilts as risk free. The premium is traditionally fixed at between 0.5 and 2%.

3.3.2 Quality of information

Information regarding property transactions is often incomplete and may not be representative or up to date. Most is subject to interpretation before application to the property under consideration. Under these circumstances, it is unlikely that there will be agreement on the value of each component.

Various studies have shown that a valuation within 5% of the control valuation is quite difficult to achieve, whilst the courts have indicated that a margin of error of 10% (and more in some cases) is within acceptable limits in considering the question of negligence. Valuers rarely attempt to use intervals of less than one quarter of one per cent when carrying out a valuation of premises let at a market rent using an all-risk yield. Using a yield of 5%, that interval itself represents a range of 5% in the resultant capital value. It should therefore be appreciated that any valuation is a considered opinion within an acceptable range. In the hands of an experienced and competent valuer, there will be a tendency for the various approximations to cancel out overall.

3.4 EVALUATION

The information gathered is used to measure performance although there is no unanimity in the conventions employed and use of information derived from a particular index should be based on an understanding of its construction. In particular, it should be borne in mind that the measures were originally developed for assessing the performance of stocks and shares and that the evaluation of property investments is different and more complicated, whilst some of the components are more subjective.

3.4.1 Range of performance measures

(a) Income growth

Income growth is the expectation of rental value appreciation, although the fact that any increase will have to await implementation in line with the provisions of the lease must be taken into account.

(b) Capital value growth

Change in capital value is a function of the combined effect of rental change and yield movement.

(c) Rate of return

Calculations are based on equivalent yield (internal rate of return assuming no rental value growth) usually adopting quarterly rests for income receipts and capital expenditure and annual rests for capital value.

(d) Total return

The total return is derived from the return on rent or income together with the overall increase in the capital value.

(e) Weighting

Weighting is sometimes employed to compensate for uneven holdings of particular types of property. Aggregation without weighting leads to the effective weighting of data by value.

(f) Chain linking

It is usual to chain link successive annual periods to provide a time weighted series of results. Given that property is a long-term investment, five years is probably the shortest period over which the performance of an individual property should be judged.

3.5 THE ACHIEVED RETURNS

There are several approaches to the measurement of the internal rate of return, the two most widely used being the time and money weighted bases. Both types of calculation have their uses but whereas the money-weighted rate is an absolute measure employed to assess the return from a particular property, the time-weighted rate is a relative measure, useful in comparing performance against that of alternative investments.

3.5.1 Weighted rates of return

(a) Money-weighted rate of return (MWRR)

The money-weighted rate of return is the discount rate which equates the sums of the initial outlay and the net income flows to the disposal or redemption value, where it is assumed that all net income is reinvested at the same discount rate. The calculation is sensitive to the timing and size of new investments which may lead to misleading results and renders the results unsuitable for comparison purposes. It is normally necessary to find the rate by iteration.

(b) Time-weighted rate of return (TWRR)

The time-weighted rate of return reflects inflow and outflow, and calculates the unique rate for the elapsed time between each cash flow, after which the results are chain linked and smoothed to arrive at an average or geometric mean representing an overall time-weighted return.

3.6   ANALYSIS

Where a portfolio has been measured to show past returns, it may be useful to disaggregate the information to relate to sectors, geographical regions or individual properties, according to needs.

3.6.1   Application

The information collected and analysed is used both to measure performance in an absolute sense and against market performance to provide a relative assessment of how well the particular property or fund performed. The information is also used in decision making, noting and reinforcing the positive aspects of performance and attempting to avoid past mistakes so as to improve future performance, providing a benchmark for which to aim.

Performance measurement is an important management tool the use of which is likely to be increased in the future. However, given the complexities and subjective aspects of even the best indices, it would be wrong to expect too much of the information provided. In particular, it should be noted that capital value increases may be more significant than rental growth in determining performance. The information should therefore be treated with care, since property cannot be traded immediately as is the case with stocks and shares: capital values may fall as well as rise with a consequent effect on any performance calculation.

FURTHER READING

Lumby, S. (1994) *Investment Appraisal and Financial Decisions*, 5th edn, Chapman & Hall.
Contemporary property market reports and indices (where published).

The computerised office

4

The introduction of computers and customised software has been the greatest advance ever achieved in property asset management, facilitating the recording, collating, sorting, manipulation employment and dispersal of information with speed and efficiency. This chapter suggests the extent to which even small practices can take advantage of recent advances, resulting in low cost systems based on reliable hardware and software.

4.1 INTRODUCTION

At its basic level, property asset management is concerned with receipts of rent and other income, the payment of outgoings and other expenses and accounting to the owner at regular intervals. It is very easy to become enmeshed at this level in an attempt to ensure that the owner receives a remittance promptly once the rents have been collected. It is equally as important to the long-term health of the investment that the manager has time to visit the properties on a regular basis and be alert to the opportunities available from time to time for actions that improve the attractiveness of the investment.

Even partial use limited to the basic word processing facilities of a computer make a dramatic contribution to reducing the workload and ensuring that mistakes do not occur in names and addresses, descriptions and similar matters once they have been recorded correctly and stored. The addition of a spreadsheet capacity enhances the ability to show detailed financial information, apportionment of incomes, outgoings, VAT and fees as necessary together with any mathematical calculations required.

The employment of a relatively simple computer-based accounting program enables even a firm of modest size to produce accurate and clear statements, rent demands and other repetitious or derivative financial information promptly and without elaborate preparation or research once the relevant information has been stored. The information on that program together with related programs for analysis, measurement, valuation and reporting purposes will assist the property asset manager in the constant search for the most effective preservation and enhancement of the net return whilst minimising the level of risk exposure.

The generally improved appearance of the detailed information should have the effect of increasing the confidence of the recipient, as well as adding to job satisfaction.

4.2 THE USE OF COMPUTER PROGRAMS

Not many years ago, the effort required to transfer from manual records to computerised management with substantial setting up and operating costs and the attendant risk of failure proved too daunting for most but the major property management practices. There were several examples of failed projects where large practices expended considerable human and financial resources in designing a system only to find that it was unreliable or inaccessible in that it was too unwieldy and complicated to be operated by any except specially-trained staff.

Today, by contrast, there are several well-tried and widely-marketed commercial property management systems. Whilst each is capable of operating in its designed form, it is usual to customise the program to incorporate particular aspects required by an individual manager and offers the option of further facilities capable of enhancing the service provided to clients. Changes required by the user usually provide few problems in incorporation into the program. Many commercial programs have achieved significant development benefits by incorporating adaptations and improvements first requested by users as additions to the standard program. It is important that the user is comfortable with the program and that it is designed to serve rather than to control.

In any sizeable office, the information held on the system may be required by more than one user at a time; it can be accessed by any other computer screen by networking, discussed in more detail shortly.

Modern commercial software is much more user friendly than it was even four or five years ago, making it possible for the office to purchase commercially available compatible word processing, accounting, spreadsheet and database programs to develop their own packages. This route has the advantage of the experienced property asset management being involved, enabling them to specify their particular requirements. Such a route should not be undertaken lightly; the initial work is very time consuming and demands a level of computer literacy achieved only with some effort.

The generally improved appearance of the detailed information should have the effect of increasing the confidence of the recipient, as well as adding to the job satisfaction of the person producing the printouts. For the more ambitious user or for the user dedicated to a full use of computer contacts, there are many other facilities available.

4.3 FREEDOM FROM CONSTRAINTS

Computerisation offers independence in many ways not previously possible whilst the convergence of telephone and computer technologies provides the basis of a very powerful and flexible tool. All the information may be assembled, manipulated and stored in one location whilst being available through networking and telephone access to other members of staff in another part of the town at the other end of the country or across the world. The opportunity exists to locate in low cost premises in an area where there is a good supply of suitable labour. The cost of telephone links is falling in response to international competition and a large amount of information may be transferred in a short time so that expense is unlikely to be significant in the context of other savings.

4.4 AN INTEGRATED COMPUTER-BASED SYSTEM

The user of computer hardware and software must accept a requirement to gain a basic level of understanding of access, input and search. Once over the initial learning hurdle, most users come to rely heavily on the availability of material to aid their daily tasks. The range of assistance at hand has reached the stage of the 'information super highway' concept which produces a vivid mental image of a high speed track with numerous lanes and opportunities to select at will various additional services offered along the route. Unlike a physical highway, there are no perceptible speed limits or hold-ups, collisions are not serious and direction can be changed without difficulty.

Figure 4.1 indicates some of the possibilities of a reasonably sophisticated system devised for use by the property manager.

It will be seen that the functions to the left of the route are the routine or mechanical ones whilst those to the right involve the collation and analysis of management information providing support for an informed view on which to base strategic advice. The reality is less simple. It is extremely difficult to categorise information in this way; recognition of items that are in themselves quite small may nevertheless prove significant and affect the decision making process or the advice proffered to the client.

4.5 MAIN COMPONENTS OF THE SYSTEM

The following account is intended to summarise the content and use of each main grouping of information. Any commercial system will operate these and other features; there is considerable sophistication in modern reasonably-priced hardware and software systems.

4.5.1 Individual property records and information

The estate terrier has always been the basic record of the client's property holdings. Typically, in a manual system, it consisted of abbreviated information on a single sheet of paper on which was recorded a summary of all the permanent information relating to the property – owner, tenant, tenure, construction, rating assessment, use class, floor area, rent payable

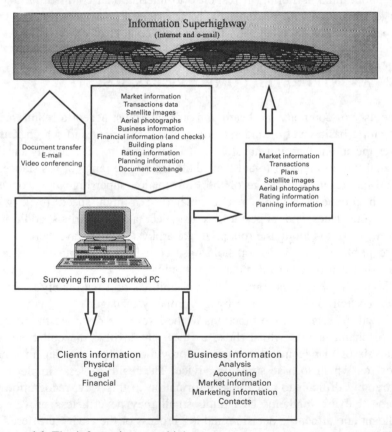

Figure 4.1 The information superhighway.

and due dates, rent review and lease termination dates, address of tenant for correspondence (if different), main terms of the lease (repairing covenants, insurance provisions, restrictions on use and assignment) and possibly a note of any unusual aspect of the property or its letting. Where a computer-based system operates, the principle is the same although access is greatly improved. Instead of relying on the terrier as a prompt, needing recourse to the files and records for further information, the program can be so arranged that the user moves from the broad information shown on the basic screen to as much detail as may be required for the particular enquiry. For example, Figure 4.2 shows the way in which more information on the basic floor area would be revealed by calling successive screens until the level of detail required was achieved.

4.5.2 Photographs, floor plans, leases, etc.

Photographs of the building may be scanned in and viewed from various angles. Proposed internal rearrangements could be indicated on the stored floor plan and the effect on lettable area, rent, service demands and other aspects ascertained.

With modern leases running to 50 or more pages and often following a particular house style as opposed to the traditional format it is helpful to be able to store a copy of the complete lease. As an aid to effective use, keywords may be specified, enabling the user to ensure that no relevant part of the lease is overlooked. For example, most leases contain a clause permitting use of the premises for certain purposes and possibly restricting any use for other purposes, whilst references to use may occur in other parts of the lease. It would be possible to specify keywords (in this case 'user') to have those parts of the lease displayed on the screen. If necessary, a printout of the relevant section or sections could be obtained as a file note or incorporated in a letter or report.

4.5.3 Accounting aspects

The system will produce rent demands and reminders, record payments, prepare periodic statements to owners and carry out any other financial aspect of the management task. With regard to service charges, not only would it record and apportion all costs incurred but it would be possible to compare the overall cost and the cost of each component against those of other buildings under management or from external indices.

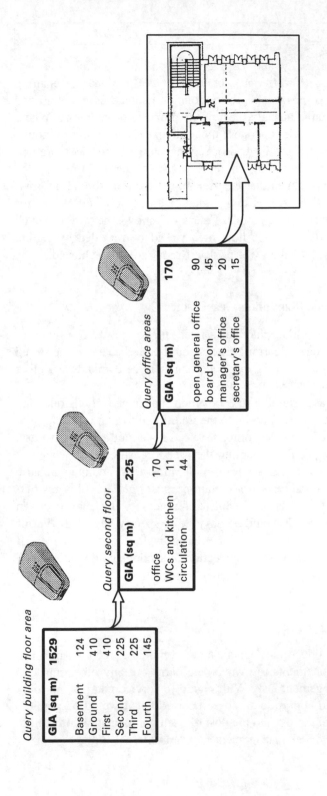

Query building floor area

GIA (sq m) **1529**

Basement	124
Ground	410
First	410
Second	225
Third	225
Fourth	145

Query second floor

GIA (sq m) **225**

office	170
WCs and kitchen	11
circulation	44

Query office areas

GIA (sq m) **170**

open general office	90
board room	45
manager's office	20
secretary's office	15

Figure 4.2 Example of program output.

4.5.4 Valuation processes

Valuation models can be stored to facilitate the production of valuations for asset, insurance and market value and to measure and monitor performance. The valuer's role is not diminished. Initially, the program may be run using default values previously supplied by the valuer and stored in the system but one has the freedom to adjust yields, growth estimates, target rates and any other factors at any time.

4.5.5 Diary function

The program will provide reminders of meetings, matters requiring attention or review, deadlines (for example, rent review trigger dates), renewals (for example insurance renewals) and similar matters.

4.5.6 Range of stored information

The program can store Ordnance Survey plans, GIS (geographical information systems), building and floor plans, manuals and images of systems demonstrations and be interfaced with external databases of which there is an increasing number. They include FOCUS (providing market intelligence), IPD (performance) and LEXIS (references to case law). It is also possible to hold rating lists (national, regional or local), telephone directories, post codes and much other information which the user judges to be worthwhile. A fully integrated installation should deal not only with the standard tasks referred to above but with space allocation, the provision and management of facilities, the financial and legal aspects of property management and the information required as the basis for performance measurement and strategic decision making. There are proposals for purchase prices for transactions dealt with by the Land Registry to be made public. It is already possible to obtain ownership details. Price information has long been available in Scotland.

4.6 EXPERT SYSTEMS

Knowledge-based interactive computer systems are being developed and it is anticipated that they have applications in the decision-making process of property investment. A short description of their ability to handle masses of complicated information and make logical deductions was provided in Chapter 2.

4.7 NETWORKING

In any sizeable office, the information held on the system may be required by more than one user at a time; it can be made available simultaneously on any other computing screen through networking, where each screen is connected to the master screen by cable and operated by the user at that point. The machines in the networked system may be at the other side of the office, in the next room, on the next floor, in a building on the other side of the town or in a more remote location. Except for those screens within cabling distance, a telephone link will be necessary but the time cost is not significant in terms of the amount of data to be transmitted. At some future stage, the cable link will be superseded by infrared and radio transmissions, overcoming the current problem of establishing links in listed buildings.

4.8 MAINTENANCE AND TRAINING

Apart from appropriate hardware and compatible software, user training is vital to enable the operator to use the facilities with understanding and to exploit the full potential of the system. It should also be appreciated that a hardware maintenance contract will be required and that software support will be accompanied by upgrading from time to time. The wellbeing of operating staff should be considered, especially in the provision of well-lit, comfortable and appropriate work stations

4.9 SECURITY

The question of passwords and security generally is a very important one and strict procedures should be laid down and enforced to protect the information stored. Levels of access may need to be considered and it should be clear who has authority to make alterations to the permanent information stored.

Not all staff need access at all levels and it would be wise to determine the access at levels appropriate to each operational group. It is essential with any program that the permanent information can be altered, amended or deleted only by authorised members of staff, when justified by a change taking place. Even authorised changes require a careful procedure to ensure that the intention is achieved.

Financial checks need to be built in to frustrate any accidental or fraudulent attempt to transfer or withdraw unauthorised sums of money whilst the system should also be used to ensure compliance with the profession's accounting regulations. It is an essential precaution that all information is duplicated and stored in another place so that in the event of sabotage, destruction by fire or other hazard the continuity of the operation is assured. Where payments are being received and made, the system should ensure that bank reconciliation and other checks preclude unauthorised payments or withdrawals.

4.10 DATA PROTECTION LEGISLATION

The Data Protection Act of 1985 requires the registration of all computer users who store information about individuals on their databases. The Act is intended to ensure that the nature and use of personal data held in computer systems is not abused and is open to scrutiny by the person concerned. Individuals may either examine a public register or enquire in writing as to what information is held about them. The Act requires that any data held legitimately is protected from unauthorised use and entitles any individual suffering damage or distress from any lapse in security to claim compensation. The Act requires data users to register with the Data Protection Registrar and renders the holding of personal data by an unregistered person a criminal offence; similarly the provision of bureau services without registration as a computer bureau is a criminal offence.

4.11 THE FUTURE

There is every indication that computer hardware and programs will become cheaper, more powerful and easier to use. That suggests that the quality of property asset management will be limited only by the ability of the individual or team formulating the advice on the basis of the information and analyses available. The opportunities offered by the emerging technologies are far from being fully utilised and exploited. The facilities now available provide an opportunity for the property manager to provide an efficient, reliable and prompt service at reasonable cost to the client.

FURTHER READING

Beddie, L.A. and Raeburn, Scott S. (1989) *An Introduction to Computer Integrated Business*, lst edn, Prentice-Hall.
Rothwell, D. (1993) *Databases: An Introduction*, McGraw-Hill.

Costs, expenses and outgoings associated with property interests

5

The costs associated with holding stocks and shares do not extend beyond those incurred for financial management and advice and any commission and stamp duty payable on acquisitions and disposals. By contrast, the various costs and charges inseparable from the ownership of property are substantial and necessary to maintain its utility. The owner's desire will be to make the end user responsible for as much of these costs as possible, thereby avoiding not only the uncertainty of the level of many of the expenses, but also the expense incurred in the detailed organisation, management and supervision of any expenditure. But even where the tenant or occupier carries responsibility for most of the expenditure there will be a need on the part of the landlord to ensure that the tenant is complying with his obligations under the lease, to monitor what is done and to check that the outcome is satisfactory.

This chapter sets out details of the main costs and outgoings, including local taxation and the ubiquitous value added tax, and looks at good practice in the physical and financial management involved.

5.1 INTRODUCTION

The right to an interest in land carries with it responsibility for a level of expense. Unlike stocks and shares which investors may, if so minded, put away and forget until they wish to dispose of them at some time in the future, even the most basic property interest requires a management input, itself incurring a charge and likely to lead to other expenses.

For example, a small plot of land near existing development and itself suitable for development may not yet be capable of development because drainage or some other essential service is not available or because the planning authority deems its development premature. Nevertheless, prudent owners will inspect the property at regular intervals to check that nothing untoward is occurring. They or their agent will check the condition of the boundary fences and ensure that no neighbour is encroaching on the land or that any other use is taking place which might lead to the establishment of a possessory title or rights in or over the land. They will wish to ensure as far as is possible that no one is acquiring rights to support or light, for example, both of which may have a limiting effect on their own later proposals. They may find that the land carries a rating assessment. They should insure against loss by fire or other hazard if appropriate and take out cover for third-party liability. A larger site and especially a site on which stands buildings or other structures will increase the need for regular inspection, proper management and judicious expenditure.

5.2 THE FUNCTION OF THE PROPERTY MANAGER

In the context of maintenance, the function of the property manager is to maintain the building to an appropriate and acceptable standard at reasonable cost and with the minimum of inconvenience to the occupier.

During the economic life of a building, the accumulated amount spent on maintenance is likely to be significant when compared with the initial or capital cost. A swift response to unscheduled failures will remain necessary though regular inspections are likely to reduce such occurrences. The technical manuals, video records and demonstrations supplied by the contractor and subcontractors will be invaluable in giving precise information and instructions to the maintenance team.

In most cases the property manager will be exercising authority conferred by the landlord and, on occasion, may find it necessary to carry out work which is the responsibility of the tenant where the tenant fails to do so and where the building is falling into disrepair as a result. In such cases, it will be helpful in any action for the recovery of expenses incurred to have detailed survey notes and photographs in addition to copies of letters and records of telephone conversations. The tenant has statutory safeguards that are considered in more detail in Chapter 6.

5.3 OUTGOINGS AND REIMBURSEMENT

The allocation, organisation and management of outgoings is an important aspect of property management and the quality of implementation is likely to have an effect on net income and capital value. It is therefore proposed to review the approach to outgoings under six broad groupings: repairs and maintenance; planned maintenance; facilities provision; insurance cover; legal and financial fees and costs and local taxation.

Owners are more interested in the net income than in the total income; where they are responsible for various outgoings they have not only to meet the cost of the work, sometimes unpredictable in terms of timing and amount, but also incur a management expense in organising and supervising the work and checking that it has been carried out in a satisfactory manner before discharging the account. The way in which the lease is structured and in particular the method of payment by the tenant is therefore of importance.

It is important that the landlord's responsibilities for outgoings are well managed and the eventual liability minimised by timeous action on the part of the managing agent.

5.3.1 Allocation of responsibility for outgoings

Where investors let properties, they will endeavour to make the tenant responsible for as many as possible of the various outgoings in addition to rent: this not only reduces the level and cost of detailed management but removes from the landlord the uncertainty of the costs involved in meeting the various obligations and therefore of the risks attaching to the investment. This is a real consideration where rents are agreed and fixed for a term, typically five years, whilst costs are incurred from job to job or, if under contract, possibly from year to year and likely to increase at least by the rate of inflation.

It is not always possible or appropriate to transfer all the cost burdens to the tenant. There are three broad categories where this applies:

1. In the majority of residential tenancies, statutory provisions limit the extent to which the landlords can avoid direct involvement: in some cases the landlord may be required to carry out improvements also.
2. Where tenants of business tenancies occupy for relatively short periods, it is likely that they will minimise their expenditure as far as possible, often resulting in the accumulation of problems requiring major work at some future date; in such cases, it is preferable for the landlord to let at an inclusive rent.
3. Finally, a tenant might be advised to refuse or restrict the liability they accept for a building or any component part which is in unsound condition and/or costly to maintain in which case the landlord would be forced to accept a lower standard of maintenance by the tenant or retain responsibility for the work.

5.3.2 Fees and costs

Landlords and their agents have attempted to establish a custom whereby the tenant pays the landlord's solicitors costs in drawing up and executing the lease and counterpart together with any stamp duty payable. The Cost of Leases Act 1958 provides that a party to a lease shall be under no obligation to pay the other party's costs unless there is an agreement in writing to do so. The act defines costs to include fees, charges, disbursements including stamp duty, expenses and remuneration.

5.3.3 Collection of tenants' contributions

In general terms, the party legally responsible for the expenditure under the terms of the lease will issue instructions for the work or service and take responsibility for payment of the subsequent account. There are four general cases where this does not apply:

1. The property is occupied by more than one tenant, each of whom contributes a proportionate part of the costs. The only practicable way may be for the landlord to issue the order, supervise the work and discharge the account and collect from each tenant a moiety of the cost. Modern multi-tenanted buildings such as office blocks are usually dealt with in this way. In such cases, a sophisticated procedure has been developed whereby each tenant pays an estimated service charge in advance with any balance due on notification: this approach is described in detail later in the chapter.
2. The tenant is responsible for the cost of insurances but is required to reimburse the landlord who wishes to retain control of the policy, not least to ensure that the cover is adequate in terms of amount and risks covered and that it does not lapse. The tenant may benefit where the landlord has a sizeable portfolio in terms of risks covered, levels of premium and attitude to claims. Insurance, too, will be described in more detail later.
3. Cases arise where tenants cannot be persuaded to honour their obligations under the lease. When other avenues have been exhausted, the landlord may have no other recourse available to him than to serve a notice under the provisions of section 146 of the Law of Property Act 1925 as described in Chapter 6, in which case the landlord should be careful not to proceed with the work until the legal requirements have been complied with and any application by the tenant for relief dealt with by the court.
4. There is one further case where the landlord may be faced with costs to repair or renovate a property neglected by a tenant without any assurance of recovering all of those costs. In the event of a dilapidations claim on the termination of the tenancy, the claim may not necessarily be sufficient to fund the total cost of the work; it is a situation to be avoided where possible by firm action based on regular

inspections and enforcement of the lease terms during the currency of the tenancy.

5.3.4   Service charges

Tenants taking leases of individual properties became accustomed to accepting responsibility, in common with the landlord and adjoining occupiers or owners, for the maintenance and repair of items such as party walls and common drains leading to the public sewer. This provision came to be applied also to work on the structure and common parts of a building where the tenant occupied only part of the building and the landlord was seeking to transfer the liability for all repairs to the various tenants. The landlord retained responsibility for the work, engaging and paying the contractor before sharing out the cost among the tenants in whatever manner was set out in the agreement. The inclusion of many more items, often requiring regular attention, led to further provisions including the specification of a method of payment of contributions at regular intervals. The service charges are now specified at length and often managed professionally; most agreements provide for the costs of management and accountancy services to be included in the charge.

(a) Purpose

Service charges are charges for those items the execution of which is, under the terms of the lease, the initial responsibility of the owner but the total cost of which is to be met by the occupier. In buildings or developments occupied by a number of tenants, the items would include not only the shared services such as toilets, heating, etc., shared areas such as lifts, staircases and malls, but also the maintenance and insurance of the building as a whole.

Figure 5.1 reproduces the format used in the management of a particular building and lists each item, a reference to the item in the schedule authorising the charge and the proportion charged to each tenant. The merit of keeping the records in this way and submitting this level of information when sending the account is that the tenant is able to follow and check the calculations.

Provision for the payment of outgoings in this way may be because it is more convenient or practicable or more efficient. It would be impracticable for each of several occupiers of an office block to decorate their own piece of external wall or for the tenant immediately below the roof to

A.N.Y. Company
Estate Management Services

Service charge schedule

Smith, Blogs and Blore

Building	**Premier House**	Tenant		Landlord's	3.19%
Location	**High Street**	Share of accommodation	79.78%	area	
		Building as share of total complex area	13.84%	Common	17.03%
		Site as share of portfolio (insurance %)	7.79%	areas	

Head of expenditure	Item	Lease reference	Total cost per annum	Tenant's share	Tenant's cost	VAT	Total tenant's cost
Structure and retained parts	Foundations	Schedule 5 Part C Clause 14.1		1.84%			
	Roofs	Schedule 5 Part C Clause 14.1		1.84%			
	Outside walls	Schedule 5 Part C Clause 14.1		1.84%			
	Structural parts	Schedule 5 Part C Clause 14.1		1.84%			
	Others	Schedule 5 Part C Clause 14.1		1.84%			
				0.00%			
				0.00%			
	External decoration	Schedule 5 Part C Clause 14.1		1.84%			
	Internal decoration	Schedule 5 Part C Clause 14.1		1.84%			
	Party Walls	Schedule 5 Part C Clause 14.2		1.84%			
	Retained part maintenance	Schedule 5 Part C Clause 14.3		1.84%			
Hot and cold water	Water to premises – rates	Schedule 5 Part C Clause 15		12.12%			
	Toilet requisites for retained parts	Schedule 5 Part C Clause 15		12.12%			
Central heating	Gas	Schedule 5 Part C Clause 16		3.73%			
		Schedule 5 Part C Clause 16		1.18%			
				88.30%			
Plant	Lifts	Schedule 5 Part C Clause 17		88.30%			
	Central heating plant	Schedule 5 Part C Clause 17		3.73%			
	Lighting (bulbs)	Schedule 5 Part C Clause 17		12.12%			
	Air conditioning	Schedule 5 Part C Clause 17		88.30%			
	Insuring	Schedule 5 Part C Clause 17		88.30%			
Pipes	Within building	Schedule 5 Part C Clause 18		88.30%			
Fire equipment	Alarms	Schedule 5 Part C Clause 19		88.30%			
	Sprinkler systems	Schedule 5 Part C Clause 19		88.30%			
	Smoke detectors	Schedule 5 Part C Clause 19		88.30%			
Cleaning	Materials	Schedule 5 Part C Clause 20		12.12%			
Fixtures and fittings	Within building	Schedule 5 Part C Clause 21		88.30%			
Windows	Cleaning exterior	Schedule 5 Part C Clause 22		12.12%			
Refuse	Collection and disposal	Schedule 5 Part C Clause 23		12.12%			
	Skips			12.12%			
	Compactors			88.30%			
	Other equipment	Schedule 5 Part C Clause 23		88.30%			
Car Park	Repair and maintenance	Schedule 5 Part C Clause 24		12.22%			
	Supervision	Schedule 5 Part C Clause 24		12.22%			
	Lighting	Schedule 5 Part C Clause 24		12.22%			
Other services	Landscaping	Schedule 5 Part C Clause 25		12.22%			
				88.30%			
Fees	Surveyors	Schedule 5 Part D Clause 26.1.1		88.30%			
	Accountants	Schedule 5 Part D Clause 26.1.1		88.30%			
	Valuation for insurance purposes	Schedule 5 Part D Clause 26.1.2		88.30%			
	Caretaking	Schedule 5 Part D Clause 26.1.3		12.12%			
	Porters			88.30%			
	Security	Schedule 5 Part D Clause 26.1.3		2.19%			
Staff	Wages	Schedule 5 Part D Clause 27		88.30%			
	Statutory contributions	Schedule 5 Part D Clause 27		88.30%			
	Provision of accommodation	Schedule 5 Part D Clause 27		88.30%			
	Uniforms/working clothes	Schedule 5 Part D Clause 27		88.30%			
	Tools/requisites	Schedule 5 Part D Clause 27		88.30%			
Contracts	Costs of entering contracts	Schedule 5 Part D Clause 28		88.30%			
Outgoings	Rates	Schedule 5 Part D Clause 29.1		12.22%			
				88.30%			
Fuel	Electricity	Schedule 5 Part D Clause 30		12.12%			
	Gas	Schedule 5 Part D Clause 30		88.30%			
Statute	Costs of complying	Schedule 5 Part D Clause 31		88.30%			
Nuisance	Costs of abating nuisance	Schedule 5 Part D Clause 32		88.30%			
Anticipated expenditure	Sinking fund	Schedule 5 Part D Clause 33		88.30%			
Insurance	Building	Schedule 6 Clause 1.1.1		0.95%			
	Heating plant	Schedule 6 Clause 1.1.1		3.73%			
	Third-party liability	Schedule 6 Clause 1.1.2		0.95%			
Finance	Interest			88.30%			
	Commissions			88.30%			
Sub total				88.30%			
Management charges	5% of all charges	Schedule 6 Clause 1.1.5		88.30%			
						Total	

Notes

Total gross internal area of complex	13381.90 Sqm
Gross internal area of subject premises	1851.59 Sqm
Share attributable to subject premises	13.84%

Figure 5.1 Example of a service charge schedule.

collect a proportion of any cost of repair from all the other tenants in the building. The landlord may have a maintenance force in his employ and it is likely to be quicker and more efficient for them to undertake the work initially. Not least in importance is the landlord's involvement in specifying and supervising the work and ensuring good quality.

(b) Apportionment of costs

Where the lease provides for work to be carried out or expenses to be incurred by the landlord on behalf of the tenant, such costs need to be apportioned where there is more than one occupier responsible for the costs. In the case of a shop unit within a shopping centre, the recoverable expenses would probably be stipulated by reference to the types of expenditure envisaged. Among these would be included supervision and management of the centre, the cost of employing management staff, their office and storage accommodation, telephones, rates and the costs of heating, lighting, repairing, decorating and cleaning and marketing and promotional work.

The lease will lay down the basis on which the costs are to be apportioned: the most common method is on the basis of floor area where the charge is that proportion of the total costs which the floor area of the unit bears to the overall floor area. An equitable basis would provide for a differential rate in respect of common parts and ancillary accommodation such as storage, basement areas or garaging. In some cases, each tenant pays an equal proportion of the costs of services attributable to the common areas but care is needed where the tenants have different uses for the premises.

Apportionment by reference to rent paid is likely to be less satisfactory, especially where the same building carries permission for a variety of uses and therefore has a range of rental values. Similarly, the use of rateable values is unsatisfactory for similar reasons as well as being objectionable on the ground that the basis of calculation is outside the control of the interested parties and may change differentially from time to time. The service charge may include sums to be placed in a contingency or reserve fund to provide for periodic replacement or renewal of more expensive expendable items.

The future liability element becomes significant where the remaining life of the component or piece of equipment is less than, say, 25 years, depending to some extent on the discount rate adopted. Surpluses retained in reserve funds are subject to tax unless elaborate provisions are made and do not lend themselves to optimum investment as they must be available for the purpose for which they were collected in the first place. Where a tenant contributes to a sinking fund he may wish to have some assurances about the continued availability of the fund in the event of a sale. Because of these and similar complications formal reserve funds are not popular although the property manager will seek to smooth recurrent costs of such items as external decoration by determining an annual contribution based on an estimate of the cost.

Unless part of the service charge is payable in advance, the landlord may be forced to provide substantial sums to meet running costs pending receipt of payment by the tenants. Many leases stipulate that an estimated amount should be paid in advance, any balance to be paid when expenditure is proved by the production of audited accounts. Such a practice should avoid stipulating a particular sum is specified as it is too inflexible. The more usual provision is for an amount based on the previous year's actual costs uplifted by the rate of inflation or similar indicator.

(c) Statutory intervention

An increase in the number of flats, both purpose-built and converted, held on long leases has resulted in the imposition of service charges on a larger scale than hitherto.

Undoubtedly, some landlords took advantage of strict covenants drawn in their favour to present bills for charges which could neither be substantiated nor challenged. Legislation to combat the abuses and to regularise procedures is contained in the Housing Act 1985. The Act defines a service charge as a sum payable as part of or in addition to rent directly or indirectly for services, repairs and maintenance insurance and the landlord's costs of management. The relevant costs are the actual or estimated costs incurred or to be incurred to be taken into account only in so far as the costs and the standard of services are reasonable. Where service charges are paid before the relevant costs are incurred, there is provision for adjustment when the costs are known.

The provisions are discussed in more detail in Chapter 9.

(d) Fees – responsibility for payment

In calculating the service charge payable by each tenant, charges will be incurred in the form of surveyors', accountants' and auditors' fees and possibly of solicitors' fees also. These charges are normally apportioned among the tenants as part of the service charge.

5.4 REPAIRS AND MAINTENANCE

The parties will negotiate and eventually determine the terms on which the letting is to proceed.

The landlord will seek to make the tenant responsible for all repairs which is the accepted norm in modern properties. Where the property is old or dilapidated, the tenant may not be prepared to accept such an onerous liability. The parties will then negotiate to achieve a solution acceptable to both but the landlord will be very anxious to delegate as much of the repair and maintenance of the structure, fixtures, plant and equipment as is possible. The long lease structure with a current term of 25 years plays its part in persuading tenants that their interest is sufficiently large to accept a heavy liability for what is often an exercise in delaying the natural ageing process. Many institutional landlords would not enter into a letting where there was any residual responsibility for repairs (or other outgoings for that matter). At the same time, and to avoid problems at a future date, landlords must be realistic about their tenants' intentions and financial abilities. Where tenants accept major repairing obligations without carrying them out, or carry them out in an unsatisfactory manner, they will cause the landlord a good deal of trouble and it may not be possible to have the work done or to recover the cost. If it is accepted that rent is agreed at a lower level where the tenant is responsible for significant outgoings, the cost to the landlord where the tenant neglects to do the work is twofold: the margin allowed for regular work is not spent and remains to be done at the end of the tenancy.

5.4.1 Condition on entry

A schedule of state of repair and condition should be prepared prior to the commencement of a tenancy by the landlord's agent and agreed with the tenant or their agent unless the tenant has accepted full repairing liabilities. It serves as a historical record and assists in the resolution of differences of opinion. It may also provide a firm basis for a claim for dilapidations on the expiration of a tenancy.

5.4.2 Inspection and notice to repair

It is usual for a lease to contain a covenant by the tenant to permit the landlord personally or by an agent and with or without workers or others to enter at any time during the term at reasonable times after giving notice in writing to view the state and condition of the premises. Provision is also made for service of notice on the lessee to remedy any defects found as a result of such inspection and for the work to be carried out by the landlord at the tenant's expense where the tenant fails to do so.

For this purpose, a right of entry is reserved and provision made for recovery from the tenant of the costs of carrying out the work together with any charges incurred.

Agents should make use of the power to inspect to carry out regular inspection as part of their routine management. Adequate records should be maintained to enable the agent to quote the date of inspection, the condition in which the building was found and the action taken together with the outcome.

Such information improves the quality of management, instils confidence in the landlord and is often crucial where action has to be taken against the tenant.

5.4.3 Forfeiture and relief

The landlord may choose to exercise a right of re-entry reserved in the lease by serving an appropriate notice on the tenant. Section 146 of the Law of Property Act 1925 regulates the procedure and enables the tenant to apply to the court for relief from forfeiture (Chapter 6).

Most leases contain a provision enabling the landlord (and/or his agents whether or not accompanied by workers) to inspect the premises usually by giving a short period of notice in writing such as 24 or 48 hours. This is a general provision: many leases also provide for the landlord's surveyor to view for dilapidations under a separate provision requiring the tenant to pay for the surveyor's fee incurred. Where the premises are let on full repairing terms, the purpose of which is to make the tenant responsible for any want of repair, there is no need for a record of the state of repair of the premises at the commencement of the tenancy. The tenant is responsible not only for maintaining the premises in the required state of repair but first for putting them in that state should they be deficient at the commencement of the tenancy as discussed in more detail in Chapter 6. In all other cases, where the tenant's responsibility for repair and maintenance is partial, it is important for a schedule of state of repair and condition to be prepared and agreed between the parties before the commencement of the lease as the basis of the liability of the parties. On termination of the lease, the landlord's agent will inspect the premises and, where appropriate, prepare a schedule of dilapidations and wants of repair as the basis of the landlord's claim against the tenant. The preparation and resolution of interim and terminal claims will be the responsibility of the property manager.

5.4.4 The definition of maintenance

The definition of 'maintenance' offered by the Committee on Building Maintenance of the Department of the Environment, contained in its report published in 1972, was based on a definition contained in BS 3811:1964.

> Building maintenance is work undertaken in order to keep, restore or improve every facility i.e. every part of a building, its services and surrounds, to a currently accepted standard and to sustain the utility and value of the facility.

The committee added the word 'improve' to reflect the fact that most buildings have long life expectancies and acceptable standards of amenity and performance will rise substantially over their lifetime as a result of one or more of the following:

- statutory requirements, e.g. safety, health and welfare provisions;
- regulations of statutory undertakers;
- the need to maintain a public image;
- steps taken to maintain rental values.

The definition has been refined under BS3811:84 to read:

> ... the combination of all technical and administrative actions including supervision intended to retain an item in, or restore it to, a state in which it can perform a required function.

Some other definitions from the same source are useful in understanding the purpose of the maintenance plan and process:

- **maintenance schedule:** a comprehensive list of items and the maintenance required including the interval at which maintenance should be performed;
- **maintenance programme:** a time-based plan allocating specific maintenance tasks to specific periods;
- **maintenance planning:** deciding in advance the jobs, methods, materials, tools, machines, labour and time required and the timing of maintenance actions;
- **planned maintenance:** maintenance organised and carried out with forethought, control and the use of records to a predetermined plan;

- **preventative maintenance:** maintenance carried out at predetermined intervals according to prescribed criteria and intended to reduce the probability of failure or the degradation of the functioning of an item;
- **opportunistic maintenance:** maintenance of an item that is deferred or advanced in time when an unplanned opportunity becomes available;
- **deferred maintenance:** corrective maintenance which is not immediately initiated after a fault recognition but is delayed in accordance with given maintenance rules;
- **emergency maintenance:** the maintenance that it is necessary to put in hand immediately to avoid serious consequences.

5.5   PLANNED MAINTENANCE

Property managers are not traditionally regarded as members of the design team when a new building is planned, a surprising omission given their unique and deep knowledge of the problems of buildings in use and the reactions of tenants to different buildings and their features including the troublesome and expensive aspects often due to inherent design problems where their experience in the management of buildings and estates would be invaluable. Their advice would often improve the usefulness of the completed building and increase the flexibility and quality of use, thus enhancing the prospects of finding and keeping a tenant or tenants. They should also be able to make helpful suggestions regarding materials and layout. Such matters affect capital and rental values and the cost of maintenance procedures and may reduce the economic life of the building.

The intention of developers may play a part in design considerations. Where they are building for sale they will be anxious to present an attractive product but may be less concerned with longer-term issues of deterioration and maintenance. Their view may well be different where they intend to hold the building as a long-term investment. Then they will seek to avoid troublesome maintenance, especially where this may have an effect on rental value. Where they intend to occupy the premises they will tend to weigh the alternatives of immediate capital costs against deferred maintenance expenses and make a financial judgement.

In the case of developments of factories, warehouses and offices where funds are provided by institutional investors, there is already a measure of control directed towards producing a building which will let readily. The present element of over-supply and therefore choice enables prospective tenants to be more selective than in the past and to secure a satisfactory lease arrangement, especially where they are professionally advised. There is more incentive now than at any time since the Second World War for the owner to offer for letting a building which is attractive, sound, energy saving and as free of maintenance as is consistent with reasonable initial costs.

Shop premises where location is the critical factor will be less affected by such considerations. Poor construction has in any event been less of a problem with such buildings, no doubt because the major cost of developing a prime site is the value attributable to the land and the rental value is therefore less sensitive to an increase in the construction element of the cost.

The quality of design, the materials used and the standard of work done all combine to affect maintenance costs during the life of a building. The ideal building would be one designed in a flexible way and built to a high specification and consequently requiring the minimum attention during its life. It is unlikely that such a building would be the most economically viable: the typical building is much more likely to be a compromise between a reasonable standard of design and construction and a modest level of maintenance expenditure. Each decision made at the design stage has implications for maintenance costs and it is in the interests of owner and occupier that the correct balance is achieved.

The approach to maintenance will now be considered in more detail.

5.5.1 The maintenance plan

Every building should have a maintenance plan. In the case of a new building, the owner's manuals provided by the contractor and subcontractors will enable the building surveyor to assess the likely incidence of maintenance, repair and replacement and to assign an estimated cost to each component.

The plan should be comprehensive and systematic, encompassing both short- and medium-term considerations. The programme should be based on a sound knowledge of the building and have regard to the following.

(a) The life of the building

The length of the remaining life of a building should be estimated even though the estimate may be revised at some time in the future. Regard may be paid to the physical life or to the functional or economic life (usually a shorter period). Experience suggests that towards the end of its physical life a building may present major problems requiring 'first-aid' maintenance and that even then it will attract only marginal occupiers. Perhaps, then, it would be preferable to consider the economic life of the building (often coinciding with its functional life) which may be defined as the period during which a satisfactory tenant would be prepared to occupy the premises which would not require appreciably more maintenance than newer but otherwise similar buildings. Where the location is satisfactory, it is often feasible to extend the economic life by undertaking a major refurbishment of the building. For example, very attractive offices have been created out of well-sited period dwellings and from 18th-century warehouses. Where a building is listed, or where there are planning or legal problems, it may be essential to consider upgrading the existing building even though it would be more viable to redevelop the site.

The plant and machinery in the building (e.g. central heating installations, air conditioning and ventilation systems and lifts) are likely to have a much shorter life than the building itself.

(b) The standard to be achieved

Where the building is one which will be in great demand because of its location, design or facilities there should be no difficulty in setting and meeting an appropriately high standard of maintenance. But not all buildings fall into that category and it may be possible and advisable to set a lower standard, where only essential work is carried out. Where refurbishment is not practicable, a building may be assigned one lifespan with a good standard of maintenance followed by a further period where it is recognised that the same quality of tenant cannot be attracted and that a consequent lowering of the standard of maintenance will not deter the poorer type of tenant.

(c) The financial implications

Periodic maintenance items should be costed on an annual basis and related to the income available. Where the cash requirement for refurbishment or other major work is too great to be met from income there should be a reasonable possibility of adequate funds becoming available before the proposal is included in the plan. Such work can often be funded on the basis of a loan granted on the expectation of a considerably greater capital value of the building when the work has been completed.

(d) Responsibility for maintenance

Property asset managers may have direct control of maintenance work but more often they will have an important but only supervisory role in ensuring that the tenant performs his covenant to maintain the property. Regular inspections should be made for this purpose, if necessary by invoking the right of inspection reserved in the majority of leases.

(e) Financial policy

Property managers should be fully acquainted with their clients' policies and expectations. They are then in a position to take the initiative in devising a long-term plan with its financial implications for discussion with their clients. The objective in most cases is to maximise the return from the property. Such an objective demands an active and positive approach, although property managers should beware of placing themselves or their clients in a financial straitjacket since costs change over a period as do rental values and it is notoriously difficult to forecast either. The aim should be for the plan to be indicative and evolutionary rather than rigid or restrictive.

(f) Budgetary provision

Approval of the financial plan will greatly facilitate the periodic allocation of sufficient sums for its implementation. Clients are likely to prepare their budgets on an annual basis and may expect to allocate a similar sum for maintenance each year. Such an approach would not recognise the probability of fluctuations in expenditure from one year to the next, for example, renewal of a roof covering at twenty year intervals or external painting every five years will impose unequal demands on annual budgets. Nevertheless, there is no reason why expenditure should not be equated as far as possible by appropriate timing and spacing of the more expensive maintenance items provided that the building does not suffer as a result.

(g) Indifference levels

When carrying out repairs to an existing building the choice of solutions is usually limited and fairly straightforward. But no decision on any but the smallest amount of expenditure should be taken without reviewing the circumstances against the background of the repair history and other information available. A leaking roof requiring frequent repair may be best dealt with by renewal even though failure at that stage in its life had not been anticipated. The alternative of frequent repairs would appear from experience to be not only unsatisfactory but costly taking account of the need for the provision of scaffolding. To these considerations may be added the inconvenience and discomfort to the occupiers, the risk of damage to the structure, stock, records or equipment and the supervision cost.

(h) Costs in use

The relative financial benefits of repair versus renewal are often self-evident with no more than a cursory consideration. The building surveyor would expect to compare the alternatives by looking at the relative opportunity costs. The choice may be complicated by variations in initial costs, levels of maintenance and running costs and the anticipated economic or physical life of the material or component.

The cost of capital may be taken at a long-term commercial interest rate or at a rate determined by the organisation responsible for carrying out the work. The annual, periodic and capital costs are estimated and reduced to a common denominator to provide a direct comparison. The basic unit cost may be reduced to an annual equivalent or capital outcome from which it will be seen which option is the more expensive and a recommendation made accordingly. Most businesses find it more convenient to have an annual equivalent of costs as then the effect on profit is clear. The calculation may be made over the life of the building or of the component, whichever is the more appropriate period in the particular circumstances.

5.6 FACILITIES PROVISION

Facilities provision and management is a relatively new concept in the occupation and use of premises. Facilities management may be defined as the provision, co-ordination and detailed management of the services and elements and their maintenance, excluding those items specifically and legally assigned to some other person or group. This is a wide activity with uncertain boundaries, varying in its scope from one building and one occupier to another.

In broad terms, the provision of facilities is a housekeeping type of activity, more to do with the occupier's day to day needs than with the building or the upkeep of the fabric, but nevertheless seen as important by organisations wishing to maintain efficiency and present a positive company image to the outside world. The traditional landlord is as yet not much involved in the provision of facilities but it is only a matter of degree – they will arrange for the windows to be cleaned under a service charge arrangement but would not expect to steam clean the carpets at intervals. Further involvement would be anathema to many of the institutions but it may be that a move towards higher standards of serviced units for shorter periods would appeal to tenants and compensate the landlord for a greater level of involvement and risk by commanding a higher payment. Some would argue that provision at this level is best done by the occupiers since only they are aware of their requirements in any detail: on the other hand there is a good deal of sophistication in the plant, internal provision and equipment associated with modern buildings as well as increasingly the building itself to the extent that it is likely that specialist supervision by a professional manager would not only be more likely to maintain efficiency but also to keep up with changes and developments in technology.

5.7 INSURANCE COVER

Insurance is a mutual activity involving the collection of contributions which are together sufficient to indemnify any contributor from suffering a direct loss. The individual contributions therefore need to be realistic in amount related to the probability of loss so that the person covered is not disappointed in the event of a claim. In practice, most insurances are underwritten by commercial organisations that need to make a profit to satisfy their shareholders.

The owner of a property will wish to be in a position to reinstate the building in the event of damage or destruction. This is usually achieved by entering into a contract with an insurance company whereby, in exchange for an annual premium, the company undertakes to pay compensation for the cost of work necessary to repair or replace the building up to the full amount covered. The extent of the cover will normally be ascertained in accordance with a provision in the lease. The lease is unlikely to set the amount although it may set out the parameters to be applied to ensure that the cover is adequate to enable full reinstatement. Where there is significant underinsurance, the insured will be assumed to carry that part of the risk between the full reinstatement cost and the actual sum insured, reducing the claim to that proportion. Most proposals require the insured to certify that the amount covered will be maintained at the full reinstatement cost.

5.7.1 Extent of cover

The buildings may be insured against damage by fire only or for other risks also, such as storm and tempest, flood, impact, earthquake and subsidence. Extensions are available to cover third party and public liability. It is usual to include in the cover a sum for demolition of unsafe structures, site clearance, local authority charges for planning applications and building regulation submissions and for compliance with local authority regulations introduced since the building was constructed, professional fees for building services and loss of rent for a stipulated period. A business tenant would probably wish to insure for additional items such as plate glass, consequential loss, the effect of business interruption and loss of or damage to fixtures, fittings and stock and where they are paying for or reimbursing the landlord for the premium would expect to have their name and interest endorsed on the policy.

The potential for massive claims in the event of urban activity by terrorists led to the decision of the insurance industry to withdraw cover for this risk. In the short term, the government adopted the role of reinsurer of last resort. Cover in respect of terrorism is now normally provided in three layers:

1. the main policy continues to provide limited cover, usually up to £100 000;
2. the mutual Pool–Re (a pool of contributions from insured and insurers, supported by additional premiums paid direct to the fund and not subject to the payment of commission);
3. the government making up any shortfall as insurer of last resort during the initial period with the intention of withdrawing once the fund becomes established.

The risk is assessed by allocation to zones. Zone 1 covers all London post codes and the commercial districts of Bristol, Leeds, Liverpool, Manchester, Cardiff, Edinburgh and Glasgow. Zone 2 contains all other locations although target risks in those areas will be charged at zone 1 rates. Target risks in zone 1 will attract a surcharge of 50%.

5.7.2 Sum insured

There is a need for the cost of reinstatement to be assessed and for adequate allowance to be made for rises in building costs during the building interval and in the preceding period when plans are being drawn and permissions obtained. The incidence of VAT as it affects the insured should also be considered. Where a property is let as an investment, there will be a loss of income that should be insured, including any increase falling due during the period of reinstatement. Any particular requirements of the insurance company should be noted. Where buildings that are nearing the end of their useful life or that have been adapted and would not be rebuilt in the same way are being assessed, the insurance company may be willing to insure on the basis of replacement by a suitable alternative structure rather than a replacement, enabling it to charge a lower premium.

Insurance cover should normally be for the full estimated cost of replacement. In addition to the cost of rebuilding, other charges incurred would include supporting and keeping watertight adjoining buildings exposed by the damage, demolition of unsafe portions, compliance with additional requirements of the planning authority or of building regulations and loss of rent for the period of rebuilding. In times of rising costs, it is prudent to reconsider the adequacy of cover every year: many insurance companies now provide for automatic increases by means of 'indexing' whereby the cover is adjusted annually by reference to an index of building costs. Where a property is insufficiently covered the insured are regarded as their own insurer for the shortfall: not only are total losses affected but a claim for partial loss would be scaled down also.

The property manager should consult the lease for provisions relating to the insurance of tenant's improvements and should ensure that adequate cover is maintained where the landlord is responsible for effecting insurance or where he is able to stipulate the sum for which the tenant should insure. Unless the building is totally destroyed, value added tax is likely to be payable on the rebuilding or repair work and it would be prudent for the property manager to consult the owner as to whether an additional sum should be included for this contingency.

Older buildings may cost considerably more to repair or rebuild than their modern counterparts and insurance on a replacement basis may be prohibitive. Examples are multi-storey mills and many churches: in such cases it is possible to negotiate cover assuming replacement with a modern equivalent subject to safeguards for both parties generally and in particular where there is a partial loss only. Even so, the insurance premium will be relatively large and possibly in excess of the rental value. Third-party liability is necessary to protect the landlord against public liability or claims made as a result of their duties, if any, under specific enactments, e.g. the Occupier's Liability Acts 1957 and 1984 and the Defective Premises Act 1972.

5.7.3 Premiums

It is common practice for leases to provide for tenants to be responsible for insuring or paying for the insurance of the building that they occupy. The payment is usually referred to as an insurance 'rent', enabling the landlord to distrain for unpaid sums, a useful safeguard.

5.8 LOCAL TAXATION AND VALUE ADDED TAX

There follows a brief resume of the non-business rate and council tax, together with a short account of value added tax in relation to property and property services.

5.8.1 Local taxation

Rateable values were assigned to both domestic and non-domestic property prior to April 1990 and a general rate levied by the local council. Since that time, the two are treated quite differently.

(a) Non-domestic property

Non-domestic or commercial property continues to have a rateable value set by the valuation officer at 'an amount equal to the rent at which it is estimated the hereditament might reasonably be expected to let from year to year if the tenant undertook to pay all the tenant's rates and taxes and to bear the cost of the repairs and insurances and other expenses (if any) necessary to maintain the hereditament in a state to command the rent' – related to values pertaining at 1 April 1988 but reflecting the property and its surroundings existing on 1 April 1990. A new valuation list has been prepared to take effect from 1 April 1995 and thereafter at five-year intervals in accordance with the provisions of the Local Government Finance Act 1988. The rateable value is the basis of the amount payable – the national non-domestic or business rate. The introduction of the word 'national' is significant; for the first time, the rate is set, not by the council in whose district the property is located, but by central government at a standard rate in the pound throughout England (with the exception of the City of London) and with a different figure for Wales. The current charges in the pound (1994/5) are 43.2p (England), 40p (City of London) and 39p (Wales). Transitional arrangements are in place to restrict the annual increase in liability by phasing in over not less than five years. Those properties enjoying a decrease in liability are subject to a 'floor', being a prescribed percentage beyond which the charge cannot be reduced.

Subject to certain limited exceptions, the empty property rate is levied at 50% of the occupied rate liability after three months of becoming unoccupied. Where a property has never been occupied, the three-month period begins from the date of an entry in the valuation list by the valuation officer or by the service of a completion notice where either the property is complete or could be so within three months.

Some commercial properties include an element of living accommodation; only the non-domestic part of such properties, known as 'composites', is rated, the remainder being charged to council tax.

Section 45(5) of the Local Government Finance Act 1988 provides mandatory charitable relief of 80% to properties occupied by charities and used wholly or mainly for charitable purposes. Where the property is owned by a charity but is vacant, a further relief of 50% is given, provided that the property has been previously occupied by the charity. The rating authority may give further, discretionary relief under the provisions of section 47.

There are provisions for alterations to the rating list and for appeals against the assessment. Unless the appeal is resolved by negotiation, it is heard by the local rating valuation tribunal with a right of appeal to the Lands Tribunal.

(b) Domestic property

The community charge or poll tax was replaced as from 1 April 1992 by the council tax. All domestic properties were externally inspected and assessed in one of six bands relating to the estimated capital value of the property on 1 April 1991.

The bands and their range are shown in Figure 5.2. It will be noted that the bands for Wales are not the same as those applying in England.

The rate in the pound is set by the charging authority. Any unit occupied by a single occupier is charged at 75% of the standard amount and vacant properties for a maximum of six months, at 50%.

5.8.2 Value added tax

Value added tax (VAT) was introduced in 1973 as a tax on goods and services. Anything which is not a supply of goods but is done for a consideration is a supply of services. As a result of later changes, it is now payable on some transactions of a capital nature also.

	England exceeding	not exceeding	Wales exceeding	not exceeding
A	–	40 000	–	30 000
B	40 000	52 000	30 000	39 000
C	52 000	68 000	39 000	51 000
D	68 000	88 000	51 000	66 000
E	88 000	120 000	66 000	90 000
F	120 000	160 000	90 000	120 000
G	160 000	320 000	120 000	240 000
H	320 000	–	240 000	–

Figure 5.2 Valuation bands for council tax purposes in England and Wales.

The supply may be chargeable at the current rate of 17.5%, be zero rated or exempt. Where a supply is taxable or zero rated, suppliers may recover any input tax which they have paid even when, as in the case of zero-rated supplies, no output tax is charged to the recipient or consumer. Less favourable treatment is received by the exempt category because again no output tax is charged but in this case no input tax may be recovered.

Those who supplies goods and services must be registered except where their turnover does not exceed a prescribed figure in which case they may register but are not compelled to do so.

VAT is payable on the first purchase of a new commercial building. Where a site is acquired and then developed as a separate contract, VAT is payable only on the construction costs and not on the land unless the vendor is registered for and charges VAT. Where a person or entity constructs, commissions or finances a building which on the first occasion during a ten-year period is put to an exempt use, either by self-occupation or the grant of an exempt lease or licence to another, a self supply charge is triggered. As a result, VAT will be payable on the full value of the development (that is the construction and land costs). This rule has hit VAT-exempt users such as banks, building societies, hospitals, schools and charities particularly hard. It has recently been announced that the self supply charge is to be abolished. At the same time, the government has introduced a statutory concession whereby buildings not previously in residential use and converted to dwelling houses will be zero rated for VAT purposes.

The European Court of Justice has ruled that the surrender of a non-domestic leasehold interest is an exempt supply unless the tenant has elected to waive the exemption. The payment of a reverse premium remains a standard rated supply. VAT will normally be payable on service charges including any provision of staff services.

Maintenance and repair work are liable to VAT as are services provided to part of a building occupied by a tenant and charged for separately. Where the payment is for upkeep of the building as a whole it may be exempt but if it is a service charge which can be treated in the same way as rent it may be zero-rated. Fees for management services are taxable as are sporting and similar rights.

Where fire damage has occurred, reinstatement of the original building is liable to VAT whereas a new building is zero-rated as described above. When advising on the level of insurance cover the property manager faces a dilemma which is discussed in the section on outgoings.

The property manager needs to exercise care in acting for overseas clients in order to avoid becoming personally liable for the landlord's output VAT (and the same caution applies to income tax).

5.9 GRANTS AND OTHER FORMS OF ASSISTANCE

Various grants, loans and other forms of support are available from central and local government and from government-sponsored bodies such as City Challenge, Housing Action Trusts, Urban Development Corporations and Rural Development Challenge.

The government has published proposals for a single Regeneration Budget for the financial years 1994/5 and 1995/6 where a total of £1300m is available to support projects lasting from one to seven years. This is in addition to various public and private partnerships involving local authorities, training and enterprise councils and similar groupings.

An important aspect of all these initiatives is the attraction of private sector funds to share the cost; funds which would not be spent in the area or on that particular initiative without priming from the grant funds. Substantial amounts of funding are available through the European Commission. European Structural funds are likely to provide not less than £1bn each year until 1999 to the United Kingdom; other funds from the same source include the European Social Fund, the European Regional Development Fund and the Poverty Programme. Again, the importance of unlocking private sector or other funding is emphasised.

FURTHER READING

Ryde on Rating and the Council Tax, Roots GRG (general editor) 1994 (with updating service) Butterworths.

Plimmer, Frances (1987) *Rating Valuation*, Longmans.

Sherriff, G. (1989) *Service Charges in Leases*, Waterlow.

The law of real property is a vast technical area of considerable complexity and one in which property managers cannot expect, or be expected, to be experts.

Nevertheless almost any action they take in managing a client's property interests will have an effect and they need sufficient understanding both to validate their actions and to alert them to a situation or a potential situation where advice and guidance from elsewhere is necessary. This chapter emphasises those practical aspects of law most likely to exercise property managers and elaborates on some areas of importance to them.

6.1 INTRODUCTION

Prior to 1926, there were three complicated branches of law dealing with various aspects of real property. The provisions were long overdue for simplification and a committee was appointed with instructions to consider the existing position and to advise what action should be taken to facilitate and cheapen the transfer of land. Fundamental changes were introduced in land law at the beginning of 1926: whilst the branches of the law previously dealing with various aspects of real property were combined and rationalised there is a degree of subtlety in the legislation and care is needed in the transfer of an interest in land to ensure that the required result is achieved.

These major structural alterations in the law should be distinguished from judicial development of legal principles and from legislative change designed to modify certain aspects of contracts freely entered into, where the intention has been to balance the negotiating strengths of the parties (for example in the case of business tenancies) or to respond to current pressures and notions of fairness (for example, enfranchisement of long leases). Elsewhere interference has been overtly political, as in the case of rent restriction – first introduced as a temporary wartime measure in 1915 – with the result that the provision of private rented accommodation is now regarded by many investors as a risky business activity. Other changes have been deemed necessary to reflect changes in social outlook.

The grant of an interest in land in exchange for periodic payment of a rent is a well-developed commercial enterprise, enabling grantors to receive a return on their capital while continuing to exercise a degree of security and control over their investment.

Property managers are centrally involved in various aspects of the conditional transfer and subsequent supervision of limited interests in land. Once the parties have been brought together they negotiate the terms of the agreement between them and are responsible for their day to day interpretation and implementation during the currency of the arrangement and may be involved at the end of the period on questions such as the status of fixtures and fittings, dilapidations, compensation for improvements and finding another tenant or assisting in disposing of their client's interest. The purpose of this chapter is to provide a framework of understanding of the incidence of occupation so as to enable property managers to play their important role while at the same time recognising the limitations of their appointment and background, combined with the need to seek specialist legal, financial or other advice where the situation is beyond their area of knowledge or competence.

6.2 ESTATES AND INTERESTS IN LAND

There are only two legal estates in land, the freehold, or perpetual, estate and the leasehold estate, where the land is held of another for a finite period which may be of short or very substantial duration.

6.2.1 The freehold estate

The freehold estate is described in the Law of Property Act 1925 as the fee simple absolute in possession. It is in effect ownership for an infinite period (fee simple) subject only to the proviso that, since all land is technically held by the title holder as a tenant in chief of the Crown, on the death of an owner intestate and without issue the land reverts to the Crown. For all practical purposes owners have unfettered power to dispose of the land during their lifetime or by will on their death. Life interests in freehold land may be granted by a trust for sale or under a strict settlement but will last only for the life of the grantee or another named person. These two forms represent the only way in which a freehold estate may be limited in time and are used to regulate family succession in ownership.

The word 'absolute' distinguishes the grant from lesser (and relatively rare) forms whilst the term 'in possession' merely makes the distinction between rights enjoyed from when they are granted as opposed to rights to commence at some time in the future ('in remainder' or 'in reversion').

6.2.2   Leasehold interests

'Land' is defined by section 205(1) (ix) of the Law of Property Act 1925 to include 'land of any tenure, and mines and minerals, whether or not held apart from the surface, buildings or parts of buildings (whether the division is horizontal, vertical or made in any other way) and other corporeal hereditaments; also a manor, an advowson, and a rent and other incorporeal hereditaments, and an easement, right, privilege, or benefit in, over, or derived from land; but not an undivided share in land.'

It is generally understood that the estate owner has exclusive rights extending to the sky above and the depths of the earth below; such an understanding is no more than a starting point and subject to many modifications by statute and otherwise. Land may be conveyed to exclude mineral bearing ground beneath the surface; by statute an owner's rights to coal, gold, oil or water are either limited or expropriated by the state. Anything may be excluded by agreement between the parties although the occupier of the surface should have in mind the risk to the stability of their holding where they do not own or control the substrata. It will be seen from the definition that it is possible to hold a limited interest that does not include the surface although there will be a need to reach the land; for example, caves exploited commercially may well be sold separately from the land and buildings on the surface some distance above the caves. It is possible to have rights to an upper floor or part of a building without any ownership of the parts of the building above or below that level (again rights of access will be required). It is impracticable to have exclusive rights to the airspace; by the provisions of the Civil Aviation Act 1982 there is no trespass where the flight of an aircraft is at a reasonable height above ground subject to compliance with all the regulations which includes an absolute liability for all damage. In general, the nearer to ground level the more proprietorial rights are enjoyed by the owner. Branches of a tree overhanging neighbours' land constitute a trespass and may be lopped off by them, although they acquire no right to either the branches or any produce. It has been held that an owner can prevent a crane from oversailing his land, which is a trespass of the airspace, thereby gaining an economic advantage (*John Trenberth Ltd* v. *National Westminster Bank Ltd* (1979)).

A corporeal hereditament includes not only the earth but also things attached thereto. It follows that buildings erected by a tenant become the property of the owner and not only to pass to him at the expiration of the tenant's interest but may serve to increase the tenant's rent on review (*Ponsford* v. *HMS Aerosols Ltd* (1978)). There is a statutory right to compensation for improvements for tenants of business premises subject to compliance with the rules under the Landlord and Tenant Act 1927 and unless the tenants were under an obligation to carry out the work as a condition of their lease. The definition also embraces fixtures such as trees and shrubs, walls and fences. In this context, fixtures are regarded as that class of object which when physically attached to the land appears to become part of it and distinguished from chattels which whilst resting on the land do not form part of it. The tests of purpose and degree of annexation are important but not always conclusive; the distinction is important when an interest changes hands when subject to any reservation to the contrary, fixtures pass with the land.

6.2.3 Proprietary interests in land

A lease confers a legal estate subject to it being created in accordance with the law and satisfying the statutory definition of a 'term of years' absolute in section 205(1) (xxvii) of the Law of Property Act 1925:

> . . . a term of years (taking effect either in possession or in reversion whether or not at a rent) with or without impeachment for waste, subject or not to another legal estate, and either certain or liable to determination by notice, re-entry, operation of law, or by a provision for cesser on redemption, or in any other event (other than the dropping of a life, or the determination of a determinable life interest); does not include any term of years determinable with a life or lives or with the cesser of a determinable life interest, nor, if created after the commencement of this Act, a term of years which is not expressed to take effect in possession within twenty-one years after the creation thereof where required by this Act to take effect within that period; and in the definition the expression 'term of years' includes a term for less than a year, or for a year or years and a fraction of a year or from year to year.

The term of years absolute is the only other legal estate capable of subsisting in land. It occurs as an estate limited in time, granted by a landlord to his tenant and in which both parties have express or implied rights and obligations. The landlord is not necessarily the freeholder since any tenant may, subject to any restriction in the terms of their agreement, let the whole or any part of the land to another.

The term of a lease may be a few weeks or years or for hundreds or even thousands of years, determined by the wishes of the parties and to some extent by the purpose for which the lease is granted (building, forestry and mining leases are often granted for terms up to 99 or 125 years and sometimes for substantially longer terms).

A lease will never convey the landlord's entire interests in the land because that would amount to an assignment: it will always be for a term shorter than the landlord's interest, thus reserving a reversion, which may be substantial or a matter of as little as a day or so.

Possession must be exclusive; the expression 'possession' is a legal rather than a physical notion and thus extends beyond the popular meaning of the word, to include the right to receive rents and profits arising out of the land. The landlord may reserve the right to enter, at certain times or on notice, for particular purposes. Without such a reservation, he is not entitled to enter the land.

Leases must with few exceptions be made by deed. The most important exception is of parol leases taking effect in possession for a term not exceeding three years, with or without power to the lessee to extend the term, at the best rent which can be reasonably obtained. A lease for longer than three years is not within the exception even though it may be determined within three years.

Prior to 27 September 1989, by the provisions of section 40(1) Law of Property Act 1925 no action could be brought upon any contract for the sale or other disposition of land or any interest in land unless the agreement upon which such action is brought or some memorandum or note thereof, is in writing, and signed by the party to be charged or by some other person lawfully authorised by that party. On and after that date the position is governed by the Law of Property (Miscellaneous Provisions) Act 1989 which repeals section 40 and requires writing to evidence a disposition. Section 2(1) provides that 'a contract for the sale or disposition of an interest in land can only be made in writing and only by incorporating all the terms which the parties have expressly agreed in one document, or where contracts are exchanged, in each'. The document must be signed by or on behalf of each party to the contract.

A lease for not more than three years at the best rent reasonably obtainable without taking a fine or a premium may be created without any formality, a verbal grant being sufficient although it is self evident that such an arrangement is open to difficulty if any of the terms are challenged. A periodic tenancy expressed to be for example from month to month or from year to year would qualify as a lease for a term not exceeding three years, even though it continued for longer than three years.

The Solicitors' Act 1974 provides that any unqualified person who draws or prepares a lease for reward is guilty of a criminal offence. A parol lease is an agreement under hand and therefore not affected by this restriction.

There are a number of types of tenancy or lease. The terms tenancy and lease are by and large interchangeable, both referring to a legal grant of possession for a fixed period. In practice the term lease is reserved to describe longer tenancies.

(a) Fixed term of years

At common law, a tenancy for a fixed period comes to an end at the expiration of that period but statute has introduced various restrictions on the right of the landlord to obtain possession against the wishes of their tenant.

The lease must be for a certain duration and must take effect within 21 years of the date of the lease otherwise it is void. A lease for a life or lives takes effect as a lease for 90 years determinable by notice on death of the life or lives.

A lease may contain a covenant for renewal of the lease on its expiration: provision for a renewal for a term exceeding 60 years is void.

Where the covenant for renewal of the lease includes the renewal covenant in its entirety, it will be seen that the lease would be perpetually renewable. However, The Law of Property Act 1922 provides that such a provision will operate as an agreement to grant a lease for a term of 2000 years, though the right reserved to receive any fine (premium) on renewal will be lost. The tenant alone retains a right to terminate at the end of the original term by giving at least ten days' notice in writing.

Where a lease is granted for a long term such as 99 or 125 years it may be a building lease in which case it is likely to be accompanied by a covenant on the part of the tenant to undertake some specific development of the land. In such cases, it is usual to pay an initial premium representing the major part of the value of the land for the period granted, with a nominal annual payment each year thereafter. Mining leases were often granted for a term of 99 years and building and forestry leases for 999 years. The terms have their origins in the maximum periods allowed for grants by life tenants under the Settled Land Act 1922 as exemptions to a provision otherwise limiting the grant of leases to any period not exceeding 50 years. Building leases are often for longer periods, usually 125 years, which gives comfort to lending and investing institutions that a development can be profitably renewed halfway through the term. The majority of buildings will outlive their functional life, whilst 'fashion' development such as shopping centres and developments associated with leisure are likely to require frequent major and costly refurbishments if they are to continue to compete with similar newer developments. Such lengths of term allow of assignment and subletting, enabling the original leaseholder and developer to recoup their investment in the project. Whilst in the early stages there will be a fairly small difference in the value of a leasehold and a similar freehold development, in the latter stages of the long lease, it will become increasingly difficult to find a purchaser and any price agreed will reflect not only the short term remaining but the risk of a substantial claim for dilapidations by the superior landlord at the end of the term.

(b) Periodic term of years

(i) From year to year

The term need not be fixed but may run from one period to the next (week, month, year) its distinguishing feature being that it will not come to an end at the end of the initial term but will continue until one or other of the parties gives notice.

A letting stated to be at a yearly rent is likely to be a tenancy from year to year, even though there is a provision for payment of rent monthly, quarterly or at some other interval less than a year. A tenancy from year to year is one which continues until determined by notice at the end of the first or any subsequent year and may be created by express grant or by necessary implication from the facts of occupation.

In the case of a yearly tenancy, at least one half-year's notice is required, expiring at the end of any full year of the tenancy.

(ii) For less than a year

A tenancy from week to week, month to month or other period is similar to a tenancy from year to year except that the length of notice is related to the periodic interval. The parties may provide otherwise or the tenancy may be subject to statutory provisions as to the length of notice.

A month is a calendar month unless otherwise provided by the tenancy agreement.

Except where landlord and tenant expressly agree, the length of the notice will be one full period expiring on the last day of that period.

(iii) At will

The courts are not anxious to infer a tenancy at will.

Whilst of insecure basis and obscure origins, a tenancy at will has more substance than a mere licence. The tenant at will cannot part with possession either by assignment or subletting. In some circumstances, the tenant at will may be able to acquire statutory protection giving security of tenure and eliminating the nature of the tenancy at will.

> It may be that the tenancy at will can now serve only one legal purpose and that is to protect the interests of an occupier during a period of transition.
>
> *Heslop* v. *Burns* (1974)

The question is considered in more detail in Chapter 7 in its relationship to business tenancies.

It is the lowest estate known to the law, being for no certain term and determinable at any time by either party. It will determine automatically where either party dies, assigns his interest, gives notice or does some act inconsistent with such a tenancy. A tenancy at will is not usually created expressly. The parties may agree that no rent should be paid but otherwise the landlord is entitled to payment for use and occupation. Tenants have the right to remove their goods from the premises within a reasonable time after the termination of the tenancy. Payment of rent at regular intervals will convert the tenancy into a periodic tenancy.

(iv) On sufferance

A tenant who remains in possession at the end of a fixed term without the consent of the landlord becomes a tenant on sufferance. The landlord may eject the tenant at any time and where tenants wilfully remain in possession after written notice, they may become liable to pay double the yearly value. Where rent is paid and accepted there may well be a presumption that the tenancy has become a periodic tenancy.

The difference between a tenancy at will and a tenancy on sufferance is that the latter exists without the landlord's consent. Statutory provisions enabling residential and business tenants to continue in occupation after the expiration of the term and other modifications of the common law position are discussed elsewhere.

6.3 AGREEMENTS FOR LEASES

An agreement for a lease is a contract whereby one party bind themselves to grant and the other bind themselves to accept a lease at some future date. For example, an agreement may be entered into between the parties where a building is under construction and to take a lease when the property is completed.

Where there is any doubt as to whether the contract entered into is a lease or an agreement for a lease, the court will decide the issue by considering the intention of the parties and whether the document is complete in itself and intended to be acted on (a lease) or whether it contemplates a further stage (when it is likely to be regarded as an agreement for a lease). The distinction is important. In the first place, a relaxation of the need for writing in respect of certain leases noted above does not apply to agreements for leases.

The latter are governed by the provisions of section 2(1) of the Law of Property (Miscellaneous Provisions) Act 1989 which was quoted previously (p. 103).

Secondly, an agreement for a lease followed by the tenant entering into possession and paying rent will operate only as a yearly tenancy regardless of the term contemplated by the contract between the parties. This may lead to complications in that some of the covenants incorporated in the original agreement may not be appropriate to a yearly tenancy. But it is expressly provided that failure to comply with the requirements of the Act shall not affect the right to acquire an interest in land by virtue of taking possession (section 53(c)). Enforcement of the terms of the original contract is no longer possible where there is evidence of an act of part performance. The Law Commission recommended abolition of the doctrine, which was achieved by the enactment of the Law of Property (Miscellaneous Provisions) Act 1989 (in particular section 2) which repealed section 40 of the Law of Property Act 1925. In appropriate cases, there may still be equitable remedies available to any party suffering an injustice.

6.4 LEASES

Landlord and tenant enter into a contract setting out the subject matter and the rights and duties of the parties. The contract is specifically an instrument under seal which conveys part of the landlord's interest to the tenant and which is referred to as a lease. One essential feature of a lease is that the landlord retains a reversion: in other words that he does not part with the whole of his interest. There is no prescribed form of lease and any appropriate words will be sufficient: nevertheless, most leases follow a settled pattern.

The simple purpose of a lease from the landlord's point of view is to regulate the tenant's occupation to the best advantage of the landlord, while the tenant has the benefit of knowing the precise terms – rights and obligations – of his temporary occupation.

The terms of the lease are sometimes negotiated between the parties but more often than not, one or both parties are represented by agents. When the parties are at one, their respective solicitors take over detailed negotiations of the lease terms. A small property let in its entirety is unlikely to create any major problems but the letting of a suite of offices in a large block or a shop unit in a mall of a modern shopping development will present the solicitors with problems of identity, grants and reservations of access, liability for a proportionate share of the cost of repairs, services and insurance and equality with other occupiers. Landlords endeavour to enforce a standard form of lease which, while highly desirable from a management point of view, may not be altogether satisfactory to the tenant. The high standing of a particular tenant or the state of the market may make the landlord more ready to accept changes.

Solicitors may not view the premises with which they are concerned and it is important for the property manager to point out any peculiarities or features of the building, land or rights which may require special provisions in the lease. Where a building is to be let in a number of parts, the essential terms need to be determined before the first letting takes place. This is an area where solicitor and agent can and should work together very closely to the benefit of their client and in the interests of efficient and responsible management.

It is always helpful for the property manager to provide a floor plan by means of which the significance of any requirements can be more easily appreciated. A plan attached to the lease will clarify what is being granted and possibly avoid disagreement at a later date. Many solicitors now send a copy of the draft lease to property managers as a matter of course for their comments and for confirmation that the intentions of the parties have been interpreted accurately.

There has been a movement away from the traditional form of lease where the parties and premises were identified, the terms as to commencement date, rent and duration set out, followed by covenants and provisions and listing the exceptions and reservations.

Various modern versions set out the covenants and other provisions impersonally, details of the parties, premises and terms being set out in a series of schedules. This style recognises the way in which a standard lease held on a word processor may be adapted. The length of lease has increased greatly over the past 30 or 40 years and it is not uncommon for an institutional type lease of commercial premises to run to 50 or so pages. There have been attempts both to restrict the length of lease and to reduce it to plain English but neither seems to have made much impact.

Some notes follow on practical considerations for the main provisions to be included in the lease.

6.4.1 The premises

Except where there is no difficulty in identifying the precise extent of the premises, the demise should be described and shown on a plan or plans, especially where it comprises only part of a larger building. The plan is usually provided for identification only and the written document takes precedence. Where the demise is part of a larger building there are likely to be rights and duties in relation to support, access, common facilities, services and utilities.

For the purpose of maintenance and calculation of service charges, the points of vertical and horizontal division are of practical importance. The landlord must reserve the external face of a wall if he wishes to control the tenant's control over it or to make use of it for a purpose such as locating an advertisement. Similarly, tenants are entitled to extend upwards if the roof construction and covering is included in their demise and they are not otherwise restricted from doing so.

6.4.2 Commencement date and term

The date of commencement of the lease must be fixed or be ascertainable at the time when the lease takes effect and the term must be certain.

It has been noted earlier that a lease must take effect within 21 years of the date of the instrument creating it if it is to be valid. Certain classes of owner are limited as to the length of lease to be granted but otherwise a lease may be for whatever term the parties agree. A tenant in business premises is likely to favour a long term to provide stability and protect their goodwill: the landlord, on the other hand, will be reluctant to grant such a term which will leave him progressively worse off in real terms as inflation continues. The market has developed to meet the reasonable requirements of both parties by the grant of a lease for a substantial term with provision for regular rent reviews, the most usual interval being five years although this may be influenced by the relative negotiating strengths of the parties coupled with the buoyancy or otherwise of the market at that time.

It is the time-honoured practice for a tenancy to commence or at least for rent to become due and payable on one of the usual quarter days (25 March, 24 June, 29 September and 25 December) although in certain parts of the country the traditional dates are 8 February, 9 May, 8 August and 11 November.

Many businesses prefer payments to be made quarterly on the first of the month (1 March, 1 June, 1 September and 1 December) and modern leases often reflect this preference.

The tenant may negotiate an option to renew the tenancy for a further term, usually on similar terms to the existing lease subject to renegotiation of the rent and except that the option to renew must be omitted from the new lease; otherwise the provisions of the Law of Property Act 1922 converting the grant into a lease for a term of 2000 years will take effect, as noted earlier.

An option to renew must be accompanied by provisions for determining the rent to be paid under the new lease if it is not to be void for uncertainty. The use of options to renew is much less important now that the business tenant has statutory rights to renew in the majority of cases.

The landlord or tenant may seek to include an option to break at some point before the natural determination of the lease. The landlord may wish to have the opportunity to redevelop with the timing dependent on planning permission or possession of the premises. Where the tenancy is one to which the Landlord and Tenant Act 1954 applies, the landlord is not absolved from serving a section 25 notice and proving the ground specified to the satisfaction of the court. Tenants may seek to incorporate a break clause to coincide with a rent review, enabling them to vacate the premises if the rent on review is greater than they can or wish to pay.

The provision that the amount of rent reserved by a lease must be certain caused some doubt to be expressed as to the enforceability of rent review clauses. It is now settled law that arrangements of this nature are valid provided there is some certain way in which the rent may be ascertained. The rent review provisions are often very detailed and relate the determination of the rent to hypothetical terms at odds with the reality of the lease.

6.4.3 The covenants and provisions

A covenant is an agreement under seal and may be expressed or implied, positive or negative, real or personal. Express covenants are those set out in the deed whereas implied covenants are those which the law infers from the nature of the transaction as being necessary to its proper operation.

Real covenants are those which affect the nature, quality or value of the land. Subject to certain conditions, such covenants run with the land – that is to say the burden may be enforced and the benefit enjoyed by successors in title. Personal covenants are effective between the original parties only and do not run with the land. Covenants to repair, to insure and not to carry on a particular trade have all been held to be real covenants: an example of a personal covenant is an option to purchase.

The law will imply the following covenants in the absence of any express provision on the part of the tenant:

- to pay the rent, in arrears unless otherwise provided: rent continues to be payable even if the property is destroyed unless there is express provision to the contrary;
- to pay the usual rates and taxes;
- to use the premises in a tenant-like manner;
- to deliver up possession at the end of the term in the same condition, fair wear and tear excepted.

and on the part of the landlord:

- to allow quiet enjoyment;
- not to derogate from his grant;
- to give possession;
- to pay landlord's taxes;
- in the case of a furnished house, to ensure that it is fit for human habitation at the commencement of the tenancy.

These covenants do not necessarily achieve the intentions of the parties and it is much more satisfactory to make specific provision.

The original parties to a lease are bound by its provisions for the whole of the term by virtue of the doctrine of privity of contract: this remains so even where tenants assign their lease or landlords their reversion.

Where an assignment takes place there is no privity of contract but there is privity of estate between the party entitled to the lease and the

tled to the reversion, provided that the covenant has reference to
t matter of the lease.

...ere is both privity of contract and privity of estate between the
original parties. In the absence of a contractual relationship, there is
privity of estate. Where the landlord assigns his reversion he will usually
require an express covenant of indemnity: in the case of tenants assigning
their lease an indemnity covenant is implied by section 77 of the Law of
Property Act 1925. Unlike original parties to a lease, assignees are liable
only for breaches occurring while the interest is vested in them.

A covenant may be void because it requires or permits something
which is illegal, immoral or impossible. Where it is feasible to separate
that covenant from the contract, the contract will operate without it:
otherwise the whole contract will be void.

It is now proposed to discuss some of the more usual covenants entered
into by the parties.

(a) To pay rent and other outgoings

How and when the rent should be paid are contained in an express
covenant by the tenant to pay the rent as provided. The modern tendency
is to refer to any payments due from the tenant for insurance service
charge and other liabilities as 'additional rent'. There are two advantages
to this arrangement: firstly, the landlord will be in a position to distrain on
non-payment of any of the amounts and not be limited to the true rent
payment; secondly, in the event of liquidation or bankruptcy of the tenant
rent is a prior charge on the assets of the lessee, the additional security
thus offered being another reason for the high standing of land and
buildings as an investment.

(b) To pay all rates, taxes and other outgoings

The parties are in general at liberty to agree who is to be liable for
outgoings, although this right is substantially curtailed in the case of
residential property subject to the Rent Acts.

Unless there is a specific agreement to the contrary, drainage charges
and the cost of abatement of nuisances are payable by the owner. Where
there is no agreement and the costs are recovered in the first instance from
tenants, they are entitled to deduct the amount from future rent.

Such covenants are often drawn very widely to ensure that the landlord
will not be responsible for any outgoings ('even though of a wholly novel
character' as one precedent has it). A modern response to the rating of

empty premises is to provide specifically for the tenant to pay any empty rate or rating surcharge.

Value added tax (VAT) is a new tax, having the ability to be charged on capital and revenue amounts. Some rents are subject to VAT, others not. In some circumstances, the existence of a liability for VAT on rent will adversely affect rental values and flexibility for those classes of tenant unable to reclaim the whole or part of any payments (for example, banks, insurance companies, charities and those persons not registered for VAT).

(c) User

A lease may make no mention of the use to which premises are to be put in which case the tenant has freedom to pursue any use within the law. The landlord will seek to prevent the tenant from engaging in any noxious, offensive, illegal or immoral use of the premises.

The more common position is that some provisions are made as to user. The tenant may be limited to a particular use or prohibited from engaging in a particular specified use but be free to engage in any other use. The implications of any limitation on the rental value of the premises on review should be considered in relation to the benefits to be gained by the landlord from the restriction. The intention may be to retain some control over the type of trade carried on, in that the tenant will not be able to make any change without first applying to the landlord for permission: it may be intended to benefit occupiers of a number of properties all belonging to the same landlord by restricting competition in the immediate locality. Current retail practice does not favour strict demarcations of this nature. Where the landlord carries on a business, he may wish to restrict competition from nearby premises in his ownership.

The effect of user covenants on the rental value of business premises on review during the currency of a lease and also on renewal by the court under the 1954 legislation is discussed in Chapter 7.

By section 19(3) of the Landlord and Tenant Act 1927 any covenant against the alteration of use without licence or consent shall, where no structural alteration is involved, be deemed to be subject to a proviso that no fine or sum of money in the nature of a fine shall be payable in respect of such a licence or consent. The landlord is not thereby precluded from requiring payment of a reasonable sum in respect of any damage to or diminution in the value of the premises or any neighbouring premises belonging to him and of any legal or other expenses incurred.

A lease of a dwelling-house commonly contains a covenant restricting occupation to use as a dwelling-house with a prohibition against use for any illegal or immoral purpose or any purpose likely to annoy or disturb the landlord, his tenants or the occupiers of adjoining properties.

Many modern leases seek to ensure that retail premises forming part of a shopping centre development open for trading by providing minimum or standard opening hours. The requirements should be able to accommodate late night shopping at a supermarket, five-day trading by banks and some other occupiers of retail units and the custom adopted by some florists, greengrocers, fishmongers and others of closing all day on a Monday. The effect of longer opening hours for public houses and the legalising of Sunday trading for shops under The Sunday Trading Act 1994 will no doubt receive attention in leases drawn after the legislation was passed.

In a recent case the landlord company sought to enforce a regulation made under the lease requiring all tenants to keep open for trading during normal hours. The defendant company held a lease on a small part of their store adjoining an arcade containing other shops belonging to the landlord: it resisted the requirement because the arcade attracted undesirable people. It was held that the regulation was ultra vires in relation to the lease (*Bristol and West Building Society* v. *Marks & Spencer plc* (1991)).

The trading hours provision is valuable to the landlord, particularly where a shopping development is not successful for one reason or another. From the tenant's point of view the prospect of paying not only the rent but also the trading overheads is particularly onerous where the trading position of the premises has deteriorated to the point of becoming uneconomic. The individual tenant would be unlikely to survive for long whereas the public company retailer would subsidise the cost from its overall profits.

(d) Not to assign or underlet

The tenant is usually required to undertake not to assign or underlet the whole or any part of the premises. Frequently the restriction is qualified by the addition of words such as 'without the written consent of the landlord' while a further relaxation is often granted by the use of a form of words providing that 'such consent however not to be unreasonably withheld'.

An absolute prohibition operates precisely as that, whereas a qualified prohibition is subject to section 19(1) of the Landlord and Tenant Act 1927 which provides that:

> In all leases whether made before or after the commencement of this Act containing a covenant condition or agreement against assigning, under-letting, charging or parting with the possession of demised premises or any part thereof without licence or consent, such covenant condition or agreement shall, notwithstanding any express provision to the contrary, be deemed to be subject –
>
> (i) to a proviso to the effect that such licence or consent is not to be unreasonably withheld, but this proviso does not preclude the right of the landlord to require payment of a reasonable sum in respect of any legal or other expenses incurred in connection with such licence or consent; and
>
> (ii) (where the lease is for more than forty years, and is made in consideration wholly or partially of the erection, or the substantial improvement, addition or alteration of buildings, and the lessor is not a Government department or local or public authority, or a statutory or public utility company) to a proviso to the effect that in the case of any assignment, under-letting, charging or parting with possession (whether by the holders of the lease or any under-tenant whether immediate or not) effected more than seven years before the end of the term no consent or licence shall be required, if notice in writing of the transaction is given to the lessor within six months after the transaction is effected.

Where the lease makes no reference to assignment or subletting the tenant is entitled to assign or sublet the premises. Where a tenant assigns he transfers the whole of his remaining interest in the premises to another person, known as the assignee, who takes on the role of the original tenant in respect of covenants that touch and concern the land in accordance with the doctrine of privity of estate. Such covenants are enforceable by or against them. By section 77(1) of the Law of Property Act 1925 an indemnity by the assignee is implied in favour of the assignor in relation to any future breaches of covenant.

A legal assignment must be effected by deed in accordance with the provisions of section 52(1) of the Law of Property Act 1925 regardless of the status of the original letting. An assignment not by deed operates as an equitable assignment. A subletting merely creates another tier in the landlord/tenant relationship: the new tenant is responsible to his immediate landlord who, as tenant, remains responsible to the superior landlord. The position of the tenant wishing to assign and needing the consent of his landlord, is strengthened by the Landlord and Tenant Act 1988. Under the act it is the duty of the landlord to grant the consent required within a reasonable time or to provide reasons for the withholding of consent. A breach of these provisions will give rise to a claim for damages. The statutory modifications imposed on assignment and sublettings of residential properties will be discussed in Chapter 9.

(e) To repair and maintain

The responsibility for repairs should be detailed as in the absence of express provision neither party will be fully responsible and the position will be very unsatisfactory. In certain circumstances, the landlord is unable to shift the burden of repair onto the tenant; see, for example, the repair burden placed on the landlord by section 11 of the Landlord and Tenant Act 1985 (as amended by section 116 of the Housing Act 1988) which places an obligation on the landlord to keep in reasonable repair the structure and the exterior of the premises and service installations in any residential tenancy for a term of less than seven years.

(f) Not to carry out alterations or improvements

A restriction on carrying out alterations is an important safeguard for the landlord in ensuring that the premises do not lose their identity or the structure suffer physical distress. A covenant may be absolute, in which case the landlord's refusal is final. Where the covenant is qualified to the extent that alterations are prohibited except with the consent of the landlord, it is implied by the Landlord and Tenant Act 1927 that such consent is not to be unreasonably withheld in the case of improvements. In this extended form, any alteration which is an improvement is subject to provisions contained in the Landlord and Tenant Act 1927 that such consent may not be unreasonably withheld. It is not possible to contract out of this provision. Where the landlord refuses consent in spite of this proviso, the tenant may apply to the court for permission.

The court will not grant permission to the tenant where the landlord agrees to undertake the work in consideration of a reasonable rent.

Compliance with the procedure for service of notice with plans and specifications is essential if the tenant is to establish the basis of his right to compensation for the improvement on quitting.

Improvements required by the landlord as a condition of granting a lease will not be the subject of compensation when the tenant leaves. It is common to stipulate that such improvements together with any alterations for which the landlord gives his consent are to be reinstated at the landlord's option at the end of the lease. The landlord then has the opportunity to consider whether the work is of any benefit to his reversion: it may also enable him to argue that the tenant's expenditure has not involved the landlord in any benefit and thus avoid the problem of tax on a notional premium equal to the increase in the value of his reversion.

(g) To comply with statutory notices

The usual form of the covenant is designed to place the burden of complying with legislation and statutory notices served thereunder on the tenant. Notices may affect the premises or the trade or business carried on therein. The use of premises for purposes as diverse as selling gunpowder or pets requires to be registered or licensed. Some of the more common requirements are summarised below.

Much recent legislation has been concerned with standards in buildings for the benefit of occupiers, whether as tenants or employees, and various third parties.

Occupiers' Liability Acts 1957 and 1984

This is among the more important enactments. Any occupier of premises owes a duty of care to visitors so that they will be reasonably safe in their lawful use of the premises. The landlord will be responsible as the occupier in respect of any part of the building, such as common parts, over which he retains control. He may attempt to limit or exclude his liability by agreement or other means, provided that any changes can be shown to be reasonable: he cannot however exclude or restrict liability for death or personal injury. He is not liable to trespassers except that he must not deliberately set out to inflict injury. Unless he has taken every care to exclude children as trespassers, he may have a greater liability to this category.

Offices Shops and Railway Premises Act 1963

The broad purpose of the Act is to provide a minimum standard of working accommodation for employees. Provision is made for the landlord or tenant to apply to the County Court to apportion the expenses incurred. Where the work required is contrary to provisions contained in the lease, application may be made to the County Court to make the necessary modifications to the lease.

Fire Precautions Act 1971

Subject to minor exceptions the Act provides that use of premises for the purposes of a factory, office, shop or railway premises where persons are employed requires the issue of a fire certificate. There are requirements for a current fire certificate to be held at the premises. Where the premises are in multi-occupation, the owner is responsible for compliance with the regulations.

Defective Premises Act 1972

The Act places on a landlord a duty of care to all persons who might reasonably be expected to be affected by defects in the state of the premises. The landlord is required to see that they are reasonably safe from personal injury or from damage to their property caused by a relevant defect. The duty is owed where the landlord knows of the relevant defect because he has been notified by the tenant or if he ought to have known of it in all the circumstances. A 'relevant defect' is defined as a defect arising out of a breach by the landlord of his repairing obligations.

Health and Safety at Work Act 1974

The Act made provision for the introduction of regulations designed to reduce accidents and bad working conditions. The provisions are concerned largely with the relationship between employer and employee, their respective duties and the formal framework to ensure that a statement policy is prepared and implemented.

Other legislation concerning the standard or use of property includes:

- Explosives Act 1875
- Alkali Works Regulation Act 1906
- Petroleum (Consolidation) Act 1928
- Shops Act 1950
- Fireworks Act 1951
- Food Act 1984
- Clean Air Act 1968
- Guard Dogs Act 1975

REQUEST FOR ITEM IN STORE

PLEASE COMPLETE IN CAPITAL LETTERS

Items will be available from mid day on the following working day

Surname / First Name	Borrower Number
Department	Date

JOURNAL

Title (The full title and year should be recorded as listed on the OPAC.)	
Year	Month
Volume	Part No.

Journal items cannot be removed from the LRC,

BOOK

Author	Title
Edition (if applicable)	Location No.

STAFF ONLY (If unsuccessful, please tick appropriate box)

In LRC		Incorrect reference	
No longer taken		Ceased publication	
Never taken by LRC		Missing	

690.24 CIB

658.859 DAV

333.333 ASI

333.333

- Factories Act 1961
- Town and Country Planning Acts 1947–1990
- Environmental Protection Act 1990.

(h) To insure

The essentials of the covenant to insure are that the premises are covered for the usual risks (loss or damage by fire, explosion, flood, lightning and aircraft), the cover is for the full replacement cost together with fees, the payments are maintained up to date, the receipt is produced to the landlord on demand and that in the event of loss or damage the premises are reinstated without delay and to the reasonable satisfaction of the landlord, any shortfall in insurance monies being made up by the tenant.

Alternatively, the landlord may effect insurance cover and claim the premium from the tenant as additional rent. In modern developments containing numerous occupiers, it is appropriate for the landlord to insure the whole in which case there will be no need to define for this purpose individual tenants' responsibilities with regard to parts of the structure, common passages, staircases and so on. Careful consideration must be given to additional cover required by any tenant by virtue of any special or unusual risk. Loss of rent for the estimated period of preparation for and rebuilding is usually included in the cover.

At common law the tenant is liable to pay rent even where the property has been destroyed. Specific provision should therefore be made for abatement of rent during the period of rebuilding: partial destruction should be treated on a *pro rata* basis.

(i) To pay service charges

Where the premises let to the tenant are part of a larger development such as a block of offices or a shopping centre, it is likely that charges directly and indirectly attributable to occupation will be collected by way of a service charge.

The charge will include payments in respect of insurance, the land-lord's obligations to repair, redecorate, cleanse and maintain, general and water rates and the cleansing, lighting and general maintenance of the common facilities, supervision and management of the development and the cost of compliance with statutes, rules and regulations, VAT and management and accountancy fees. The covenant may also provide for the apportionment of the cost of promotional or other work carried out at the request of a majority of the tenants for the benefit of the development as a whole. The amount of the service charge will be ascertained and certified annually by the landlord or their managing agents or accountants. Apportionment may be by means of floor area, rateable value or use or by a combination of these methods or by weighting various aspects.

Large sums may be involved in providing the various services and provision is often made for a payment in advance on account of the estimated costs for the ensuing three months.

Service charges payable in respect of residential accommodation are subject to strict statutory control as explained in Chapter 9.

(j) To surrender

The lease usually contains a specific covenant on the part of the tenant to surrender and yield up possession of the premises in good repair and condition at the end or sooner determination of the term.

(k) Costs, fees and stamp duty

The tenant may be required to pay the landlord's costs of preparing and executing the lease and counterpart together with the stamp duty payable. Such an agreement cannot be enforced unless it meets the requirements of the Cost of Leases Act 1958. The Act provides that, notwithstanding any custom to the contrary, a party to a lease shall be under no obligation to pay the other party's costs of the lease unless the parties thereto agree otherwise in writing.

Costs are defined to include fees, charges, disbursements including stamp duty, expenses and remuneration. Provision is usually made for the tenant to reimburse the landlord for all solicitors' and surveyors' fees and disbursements in connection with notices under section 146 or 147 of the Law of Property Act 1925 or on any application by the tenant for a consent or licence required under the leases.

(l) Landlord's covenants

The usual covenants by the landlord are for quiet enjoyment, to maintain and repair those parts of the demised premises for which he is responsible together with the common areas and where appropriate to insure and to apply all insurance monies to rebuilding.

(m) Provisos

The lease usually continues with a series of provisos setting out in some detail the agreement between the parties in the event of certain eventualities, the more important of which are listed below:

(i) Forfeiture and re-entry

A lease is liable to forfeiture where the tenant breaches a condition of the lease or breaks a covenant but in the latter case only where the lease contains an express provision for forfeiture.

Where the breach concerns the non-payment of rent the landlord must bring an action for recovery following termination of the tenancy. Forfeiture may be waived by any act by the landlord which is inconsistent with forfeiture. Payment into court of all arrears of rent and costs has the effect of staying proceedings and enables the tenant to apply to the court for relief which is entirely in the discretion of the court. A sub-tenant may seek relief against forfeiture of the tenant's interest. Forfeiture for breach of any other covenant is governed by the provisions of section 146 of the Law of Property Act 1925 which provides, in part:

(1) A right of re-entry or forfeiture under any proviso or stipulation in a lease for a breach of any covenant or condition in the lease shall not be enforceable, by action or otherwise, unless and until the lessor serves on the lessee a notice –

(a) specifying the particular breach complained of; and
(b) if the breach is capable of remedy, requiring the lessee to remedy in the breach; and
(c) in any case requiring the lessee to make compensation in money for the breach;
(d) and the lessee fails, within a reasonable time thereafter, to remedy the breach, if it is capable of remedy, and to make reasonable compensation in money, to the satisfaction of the lessor, for the breach.

It is provided that the section has effect notwithstanding any stipulation to the contrary.

The court has power to grant relief from forfeiture and if it does so may give a direction for costs and compensation as it thinks fit. As in the case of non-payment of rent, forfeiture may be waived by any inconsistent act of the landlord. However, in the case of continuing breaches of covenant, waiver applies only to those breaches having taken place and subsequent breaches offer the landlord a further opportunity to enforce a for-feiture.

(ii) Options

The lease may contain an option for either or both parties to determine or bring the lease to an end on a date prior to that specified in the lease. The option to determine may stand alone or be associated with a rent review clause, the intention then being to give tenants an opportunity to discontinue the lease where, for example, the new rent level is not acceptable to them or where a limited planning permission has come to an end and not been renewed.

An option on the part of the tenant to renew the lease at the expiration of the term is less frequent now that the tenant has considerable security of tenure under the Landlord and Tenant Act 1954. Nevertheless, it remains a valuable right for the tenant as it precludes the landlord serving a statutory notice and citing one of the grounds provided by the Act in support of their refusal to grant a new tenancy.

An option to renew should deal with the time and manner in which it should be exercised and set out the terms of the new lease. It may be provided that the new lease is in the same form as the existing one, save that the amount of rent payable will be subject to alteration.

It was noted earlier that the effect of repeating the covenant for renewal would be to create a perpetually renewable lease which would operate as a demise for 2000 years, an outcome that the defendant landlord in *Burnett, Marjorie Ltd* v. *Barclay* (1981) was fortunate to avoid. A lease containing an option to renew for a further seven years contained the further proviso that the lease should also contain a like covenant for renewal for a further term of seven years on the expiration of the term but the court held that this clause was not part of the covenant for renewal.

An option for renewal is an estate contract registrable as a Class C (iv) charge under section 2(4) of the Land Charges Act 1972. In the event of non-registration the option is void against an assignee for money or money's worth (*Midland Bank Trust Co. Ltd* v. *Green* (1980)).

Finally, the lease may contain an option to purchase in favour of the tenant. The option should state the conditions to be observed in its exercise and either specify a price or give instructions as to its determination. The option must be exercised as provided. Unless expressly provided to the contrary, an option will not survive termination of the lease.

Again, the option should be registered. In a recent case, it was held that an option to purchase the reversion at a valuation was ineffective because the landlords declined to appoint a valuer to agree the price with the tenants' valuer. Templeton L.J., giving the judgement of the court, said:

> We arrive at that conclusion regretfully because the option was clearly intended to be effective and was at the time thought to be effective ... Nevertheless, it seems to us that ... the parties succeeded in selecting a classically uncertain form which the court cannot assist them to operate.
>
> *Sudbrook Trading Estate Ltd* v. *Eggleton and Others* (1981)

(iii) Service of notices

Provision is often made for the service of notices, by stipulating an address to which any notice must be delivered. Where there is a subsequent change of address, service should probably be made to both addresses. The form of notice may be specified but will be overridden by forms of notice prescribed by statute.

(iv) Settlement of disputes

The most likely cause of dispute under a lease with provisions for periodic reviews of rent within the lease period is the amount of rent to be paid.

The provisions for initiating the review may be very detailed: the lease will lay down the procedure for resolving any failure of the parties to reach agreement. The common arrangement is for the appointment of an appropriate person to act as an arbitrator or as an independent valuer. The lease may provide for the President of the Royal Institution of Chartered Surveyors to make the appointment. Invariably, he or she will appoint a chartered surveyor who, in addition to the primary duty of making a determination of value, may be called upon to interpret aspects of law. The Institution has provided guidance notes for the assistance of arbitrators, independent valuers, the parties and those advising them. The notes distinguish between an arbitrator and an independent valuer and proceed to discuss the appointment, acceptance, powers and duties of the appointee, the procedure before, during and after the hearing or the receipt of written representations, the form of the Award and the treatment of fees and costs.

Arbitrators will be subject to the provisions of the Arbitration Acts 1950 and 1979 but are not liable for negligence. Independent valuers make use of their own knowledge in reaching a decision and are liable for negligence.

The case of *Belvedere Motors Ltd* v. *King* (1981) is a recent example of a charge of negligence being made by an aggrieved party against the independent valuer. The judgement by Jones J. showed a very clear understanding of the valuation issues and highlights the careful and thorough way in which the defendant went about his duties. The restatement of the applicable law is well worth reading in the full report. Further information on dispute resolution is contained in Chapter 8.

6.5 LICENCES

A licence is a personal privilege which does not confer an interest in land, does not depend on exclusive possession and simply permits the licensee to do what would otherwise be a trespass. A bare licence is one granted without valuable consideration. It cannot be assigned, may be revoked at any time and will be determined automatically if the owner of the property dies or conveys their interest to another.

The owner of the land may maintain an action of trespass against a former licensee remaining after revocation of their licence. Where there is no specific provision for notice to terminate the licence, the law will imply whatever length of time is fair and reasonable between the parties. A licence coupled with an interest in land (e.g. the right to enter the property of another and cut and remove turf) may be assigned unless otherwise agreed and in general is irrevocable during the currency of the interest. It is unlikely that a mere licensee would be held liable to repair.

In an arrangement between an oil company and the operator of a filling station owned by the company, it was held that an agreement under which the grantor retained rights of possession and control over the property were consistent only with the grant of a licence: the grantee was therefore unable to claim protection as a business tenant under the Landlord and Tenant Act 1954 (*Lear* v. *Blizzard* (1983)).

The use of licences as a means of avoiding the provisions of the Rent Acts relating to security of tenure has been thoroughly explored by landlords and others and will be considered in Chapter 9.

6.6 RENTCHARGES

A rentcharge may be defined as an annual or other periodic sum charged on or issuing out of land where the owner has no reversion in the land but has power to distrain either by express provision in the instrument creating the charge or by statute. A rent reserved by a lease or any sum payable by way of mortgage is not a rentcharge.

The Law Commission considered the effect of the creation of rentcharges and reported a widely held view that the creation of rentcharges invaded the principle of freehold tenure with the sole purpose of providing a bonus for the builder: they concluded that the system required, at the very least, radical reform. As a result a private member's bill was introduced and received government support in its passage through Parliament to become the Rentcharges Act 1977.

The Rentcharges Act was passed to prohibit the creation and provide for the extinguishment apportionment and redemption of all but very limited types of rentcharge, the only two of any practical importance being the estate rentcharge and the variable rentcharge.

6.6.1 Estate rentcharges

A rentcharge created for the purpose of:

1. making covenants to be performed by the owner of the land affected by the rentcharge enforceable by the rent owner against the owner for the time being of the land or
2. meeting, or contributing towards, the cost of the performance by the rent owner of covenants for the provision of services, the carrying out of maintenance or repairs, the effecting of insurance or the making of any payment by them for the benefit of the land affected by the rentcharge or for that or other land shall be an estate rentcharge not subject to the provisions for redemption provided it is nominal in amount or unless it represents a reasonable amount for the performance by the rent owner of any covenant referred to in (1) above.

6.6.2 Variable rentcharges

A rentcharge may be variable where the amount payable is related to some index and therefore not known in advance, or where the deed creating the rentcharge makes provision for variations to take effect at certain times in the future. In either case, the Act will not affect this class of rentcharge until it ceases to be variable, which time is to be regarded as the date on which the rentcharge became payable for the purpose of extinguishment.

6.6.3 Provision for extinguishment

The Act provides simply that every rentcharge other than those within the exclusions referred to shall be extinguished at the expiry of the period of 60 years beginning on 22 August 1977 or with the date on which the rentcharge first became payable, whichever is the later.

6.6.4 Provision for redemption

Where the rentcharge is not an estate rentcharge or one covered by the other limited exceptions, the rentchargee may apply to the Secretary of State to ascertain the sum payable and in due course to issue a redemption certificate on proof that the redemption price has been paid to the rentcharge owner or paid into court where the owner is not known.

6.6.5 Calculation of redemption price

The Act contains a formula for the calculation of the redemption price which relates the price to the current return on 2.5% consolidated stock. The formula is:

$$P = £R/Y - (R/(Y(1+Y)^n))$$

where

P = the redemption price
R = the annual rentcharge
Y = the yield expressed as a decimal fractions of 2.5% consolidated stock
n = the period in years for which the rentcharge would remain payable if it were not redeemed, any part of a year being taken as a year

The same result would be obtained by multiplying the annual rentcharge by the appropriate years' purchase on the single rate basis, using a yield of 2.5%.

There is no time limit for commencing the redemption procedure: the right is a continuing one and the time of redemption is reflected in the formula for calculation of the redemption price.

There are provisions enabling the Secretary of State to apportion a rentcharge not subject to redemption as between different owners of land or different parts of land in one ownership. Such apportionments are subject to a right of appeal to the Lands Tribunal.

Applicants for redemption or apportionment must bear their own expenses and the reasonable expenses of a mortgagee incurred in producing documents of title to the Secretary of State.

6.7 GROUND RENTS AND ENFRANCHISEMENT

A popular method of achieving development is by the grant of long terms with a requirement that the grantee is responsible for creating and maintaining an approved building on the site. Many individual properties, particularly houses and industrial premises, have been developed in this way which has also been popular for housing estates. The landowner receives a relatively small but extremely well secured annual income for each site whilst the developer initially commutes and then passes on the annual equivalent of the land value, reducing his need for development finance. With the proximity of the end of some of the earlier residential building leases and the prospect of losing occupation of the property or at best remaining but at a much increased market rent there was a flurry of political activity in an attempt to diffuse what was seen as a problem. Part I of the Landlord and Tenant Act 1954 provided that in the case of a long tenancy – defined as one for more than 21 years at a low rent (stated to be a sum less than two thirds of the rateable value) – the tenant was protected on expiry of the long term and enabled to continue in occupation subject to the rights of the landlord in certain cases to serve a notice proposing the terms of a new tenancy or providing information on the grounds on which they rely in seeking an order for possession from the court. For those intent on reforming the law on the basis that such leases, although entered into freely, were socially wrong, the provisions of the 1954 Act were no more than interim or holding provisions pending a more comprehensive and fundamental reform of the law. To this end, the Leasehold Reform Act 1967 attempted a much more radical solution by enabling an occupying tenant of a house (but not a flat) within certain rateable value limits to acquire the freehold interest at a price to be agreed on the basis of a number of assumptions or to gain an extension of the lease for a further 50 years subject to renegotiation of the ground rent payable. Certain classes of dwelling were excluded from the operation of the acts, particularly flats and houses with higher rateable values. The principle has been extended by the Leasehold Reform, Housing and Urban Development Act 1993 to include flats held on long lease and also houses previously excluded because of their rateable values. At the same time, the act introduces market value as the basis of compensation taking account of marriage value.

In all these cases the provisions ensure that the tenant may acquire the freehold on heavily discounted terms, if necessary against the wishes of the landlord. For those not wishing to acquire the freehold or unable to afford to do so, there is an alternative of an extension of the existing lease, subject to payment of a modern ground rent. The valuation rules are complex and incomplete: as a result the process has evolved mainly through the Lands Tribunal, which has developed a series of stylised valuation models using rents and yields not always derived from market evidence. The initial jurisdiction now lies with the local rent assessment committees, sitting as leasehold valuation tribunals. Appeals to the Court of Appeal and the House of Lords have explored the legal interpretation of the valuation rules. The motivation on the part of tenants has been the security offered together with the opportunity to trade the property, something that was previously extremely difficult in the case of a lease with only a few years unexpired.

6.8   EASEMENTS AND PROFITS *A PRENDRE*

6 8.1   Easements

An easement is a right in *alieno solo* (in the soil of another) attached to the ownership of land and being a right to use or restrict the use of the other land. Examples are rights of way (a right to use) and rights of support and light (rights to restrict use).

The essential characteristics of an easement are that there must be a dominant and a servient tenement and the dominant and servient owners must be different persons: the easement must benefit the dominant tenement and the right must be capable of being granted by deed.

An easement is a legal easement if acquired by express, implied or presumed grant or granted by deed for an estate in fee simple absolute in possession or a term of years absolute in which case the benefit will pass to any later purchaser of the land. Recourse may be had to the Prescription Act 1832 but only where litigation is being pursued. The Act lays down a period of enjoyment of 40 years as of right and without interruption (20 years in the case of easements of light) as absolute and indefeasible. Where the easement has been created in any other way, it will be an equitable easement, enforceable against a purchaser of the legal estate in the servient tenement only where registered as a land charge. A public right is not an easement because it can be exercised by the public at large and is not dependent upon the ownership of land.

An owner may be prepared to grant an easement in or over their land. But he may be anxious that an adjoining owner does not acquire by long user any right which might restrict the present enjoyment of future development of the land. The property manager will use whatever means are available to prevent the acquisition of easements. For example, the owner who crosses his neighbour's yard as a short cut to his orchard without legal authority should be prevented from doing so or requested to make an appropriate acknowledgement, or the owner may bring an action for trespass. Some easements, for example of light or support, are likely to be acquired in time because usually there is no practical way in which to obstruct his enjoyment and their use does not involve a trespass.

An owner has a natural right of support against his neighbour or the owner of the subsoil held apart from the surface unless excluded by agreement or statute. However, the right does not extend to a building, either by adjoining land or by an attached building. The right may be acquired by user after a period of enjoyment. Where physically connected buildings have been constructed at the same time by the same building owner, each conveyance could be expected to contain reciprocal grants and reservations.

An owner has no common law right to enter on his neighbour's land to attend to a repair on his property not capable of being reached from within his boundary. Under the provisions of the Access to Neighbouring Land Act 1992, the courts have discretionary powers to order access for certain categories of work (being defined as basic preservation works, clearance of drains and sewers and the care of trees, hedges and shrubs) where they are reasonably necessary for the preservation of the dominant land and would be substantially more difficult to carry out without entry. Access for other works may be ordered if considered fair and reasonable in all the circumstances. The order does not endure and where agreement is not possible, a new application would be required each time access is required.

6.8.2   Profits *a prendre*

A profit is, like an easement or a rentcharge, an incorporeal hereditament. It is a right to take certain limited things from the land of another (the servient tenement). It is not necessary for there to be a dominant tenement but, in most cases where there is, the right is described as a profit appurtenant. Where there is no dominant tenement it is a profit in gross.

Profits exist as the right to take something from the soil itself (e.g. sand or gravel) or the produce of the soil (e.g. to pasture one's cattle or to take fish). A profit may be acquired in similar ways to those described for easements, except that a profit in gross cannot be acquired by statutory prescription.

6.8.3   Extinguishment

Both easements and profits may be extinguished by release either express or implied: by abandonment (not merely user), by operation of law or by unity of ownership possession.

6.9   PUBLIC RIGHTS

A public right is a right enjoyed by the public at large which does not depend on ownership of land. The most common public right is a right of way which way may be created by statute or by dedication and acceptance, usually inferred.

The Highways Act 1980 provides that a right of way is deemed to have been dedicated where it has been enjoyed as of right by the public for twenty years or a shorter period where it can be shown that there was an intention to dedicate. Interruption of the use will defeat a claim. This may be physical, for example by closing the way once a year, or may be achieved by exhibiting a notice on or close to the way or by depositing a map with the local council and renewing it by means of a statutory declaration every six years.

It should be noted that the soil beneath the public right of way or highway remains vested in the adjoining owners and should the road cease to be a public highway, the land will revert to the adjoining owners. A practical example of such a case occurred when, as part of a pedestrianization scheme in a shopping street, a local authority proceeded to stop up a side street under powers contained in the Highways Act. The adjoining owners were able to combine and benefit to the extent of

£250 000 by selling the freehold in the former highway as the site for a shop development.

FURTHER READING

Burn, E. H. (1994) *Cheshire & Burn's Modern Law of Real Property*, 15th edn, Butterworths, London.
Gray, K. (1993) *Elements of Land Law*, 2nd edn, Butterworths, London.

Business tenancies

7

The purpose of this chapter is to examine the important provisions of current legislation which protect tenants of business premises and the interpretation of these provisions by the court.

The property manager plays a key role in regulating business tenancies and is unable to manage and negotiate rent reviews and renewals of leases on behalf of either landlord or tenant without a proper grasp of the underlying statutory provisions and the case law flowing therefrom.

7.1 INTRODUCTION

Nowadays it is commonplace for the law to intervene to modify a contract, in order to afford a degree of protection to one of the parties to it. But it was not at all common when the Landlord and Tenant Act was passed in 1927, its main purpose being to give the tenant a right to a new lease or to some measure of compensation where a new lease was not available.

A majority report of a government appointed Leasehold Committee discerned a widely held belief that landlord and tenant were unable to bargain from positions of equality and that the balance was heavily in favour of the landlord in spite of the intentions and provisions of the 1927 Act. The recommendations of the Committee sought to buffer the worst effects of the landlord's ability to obtain possession of premises occupied by a tenant and resulted eventually in the passing of the Landlord and Tenant Act 1954, Part II especially of which applies to business tenancies. This Act remains as the principal act though the Law of Property Act 1969 made some improvements in its operation and effect: the 1927 Act now survives only in respect of the provisions relating to compensation for improvements carried out by the tenant and this subject to modifications contained in the 1954 Act.

Some parts of the legislation have generated much litigation, especially over the last twenty-five years or so, but in general the code has worked well and proved acceptable to both landlord and tenant. Tenants have enjoyed a considerable degree of security of tenure, with provisions for compensation where they have been required to vacate the premises against their wishes under certain rights of repossession reserved to the landlord in cases where the correctness of the tenant's conduct has not been in question. At the same time, the landlord has retained freedom to include in the lease agreed provisions for the review of the level of rent payable during the currency of the lease. The success of the statutory intervention may best be judged by the high level of activity in the investment market in business premises and by the acceptance of relatively low initial returns on such an investment.

Indeed, it is doubtful whether the landlord and tenant relationship in business premises could have developed without a framework of the kind now in force: without the confidence instilled by the legislation, many tenants would have sought to avoid the risks inherent in a free market by acquiring freehold premises – perhaps then much of the major commercial and industrial development of recent years would not have taken place.

7.2 DEFINITIONS

Part II of the Landlord and Tenant Act 1954 as amended by Part I of the Law of Property Act 1969 applies '. . . to any tenancy where the property comprised in the tenancy is or includes premises which are occupied by the tenant and are so occupied for the purposes of a business carried on by him or for those and other purposes' (section 23(1)).

It is necessary to consider the several parts of this provision with help from the Act and from decisions of the courts. First, it must be emphasised that a tenant who does not occupy or a tenant who occupies but does not carry on a business, is not afforded protection under the provisions of this part of the Act.

7.2.1 Tenancy

'Tenancy' means a tenancy created by a lease, underlease, agreement for a lease or an underlease or by a tenancy agreement unless expressly excluded by the Act but excludes certain types of tenancy and does not include a mortgage term or any interest in favour of a mortgagor arising from an attornment clause.

Where a tenant went into possession on payment of three months' rent but before completion of a lease which was not completed because the tenant objected to one of the terms, namely the requirement to deposit a sum of money, he claimed the protection of the Act. On appeal it was held that a tenancy derives from a consensual arrangement and that particular circumstances can rebut the legal presumption of a tenancy: entry into possession was said to be a classic example of the way in which a tenancy at will may exist (*Javid* v. *Aqid* (1990)).

7.2.2 Premises

The courts have given a wide interpretation to the word 'premises' so as to extend to property other than buildings. For example, land let to be used as gallops was held not to be an agricultural holding (as claimed) and was not therefore excluded from the provisions of Part II of the Act (*Bracey* v. *Read* (1963)). On the other hand, it was held that a right of way was not and never could be within Part II as being occupied for the purposes of a business (*Land Reclamation Co. Ltd* v. *Basildon District Council* (1978)). Nor was an easement or something akin thereto (*Jones* v. *Christy* (1963)).

7.2.3 Occupation

Tenants cannot claim security of tenure where they have sublet the whole of the premises. Where they have sublet only part, remaining in occupation of the other part, they are entitled to security in respect of that part of which they remain in occupation. Where the landlord requires the new tenancy to comprise the whole of the property included in the current tenancy, tenants cannot seek a new tenancy only of that part which they occupy.

A flat leased to a medical school and occupied by four medical students was held to be occupied for the purposes of a business in *Groveside Properties Ltd* v. *West Medical School* (1983) on the basis of evidence to the effect that the school wished to foster a collegiate spirit.

Premises used only for the storage of files and infrequent lunches with a temporary licence of part was held in *Hancock & Willis* v. *GMS Syndicate Ltd* (1982) to be no longer occupied so as to qualify for the protection of the Act.

In *Cristina* v. *Seear* (1985) the tenants owned all the shares in a business carried on by a limited company but it was held that the premises were not occupied by the tenant for the purposes of section 23(1).

In *Linden* v. *Department of Health and Social Security* (1985) which concerned eight self-contained flats occupied by employees of NHS hospitals it was held that the authority could claim to be in occupation due to its managerial functions, holding of keys of vacant flats and that exclusive possession was not granted to the occupier. To emphasise how fine a line there is, depending on the particular circumstances, in a case where the tenants had a tenancy of and let out lock-up garages, doing some incidental works and occupying one as a store but only visiting twice weekly by staff they were held not to be entitled to protection under the Act (*Trans Britannia Properties Ltd* v. *Darby Properties Ltd* (1986).

7.2.4 Business

The expression 'business' includes a trade, profession or employment and any activity carried on by a body of persons whether corporate or unincorporate (section 23(2)). But where the tenant is carrying on a business in breach of a prohibition of use for business purposes (as distinct from a prohibition of use for the purposes of a specified business or of any but a specified business) the premises will be excluded from the provisions of the Act unless the immediate landlord or his predecessor in title has consented to the breach or the immediate landlord has acquiesced therein. The courts have had many opportunities to consider the question of what constitutes a business.

A much quoted definition is that business means '. . . almost anything which is an occupation as distinguished from a pleasure: anything which is an occupation or a duty which requires attention . . .' (Lindley L.J. in *Rolls* v. *Miller* (1884)). The term has been held to include a tenancy of a tennis club (*Addiscombe Garden Estates Ltd* v. *Crabbe* (1957)): premises let to a Minister 'for and on behalf of Her Majesty' were held to be occupied by the Crown for the purposes of a business carried on by the Crown (*Town Investments Ltd* v. *Department of the Environment* (1977)). Premises occupied to carry on a seasonal business (*Artemiou* v. *Procopiou* (1965)) and the activities of the governors of a hospital in administering the premises (*Hills (Patents) Ltd* v. *University College Hospital* (1956)) were both held to be occupied for the purposes of a business.

A subletting may be a business but it is protected only if the tenant occupies (*Bagettes* v. *GP Estates Co. Ltd* (1956)).

Premises where tenants occupied part as their residence subletting the remainder for business use and providing various services were held to be occupied for the purposes of a business and thus protected (*Lee-Verhulst (Investments) Ltd* v. *Harwood Trust* (1973)).

It is not necessary for the business to be carried on at the premises for Part II of the Act to apply to the tenancy: the requirement is that it is occupied for the purposes of a business: it is thought that a lock-up garage used to store a van used in a business would be within the provisions of the Act. But where a hotel owner took a tenancy of premises to accommodate staff employed in his hotel it was held not to be within the provisions of the Act as it was not necessary for the staff to live there and convenience was insufficient (*Chapman* v. *Freeman* (1978)).

A tenant who used premises to conduct a Sunday School for one hour each week was held not to be carrying on a business (*Abernethie* v. *A.M. and J. Kleiman* (1970)).

7.2.5 Holding

The term is applied to the premises comprised in the tenancy, excluding any part which is not occupied either by the tenant or by a person employed by him for the purposes of a business. It should be noted that the tenant does not have to occupy the whole of the premises for business purposes, but if he does so occupy any part, any other part occupied by him for other purposes (e.g. as living accommodation) is also entitled to the protection available under the Act.

7.2.6 Landlord

The description 'landlord' may be qualified by one of three adjectives – competent, mesne, superior. Each has a special meaning for the purposes. of Part II of the Act.

(a) Competent landlord

References to the landlord normally refer to the 'competent' landlord, i.e. the person who, at the time, has

1. an interest in reversion expectant (whether immediately or not) on the termination of the relevant tenancy and
2. an interest which is either the fee simple or a tenancy which will not come to an end within fourteen months and further that no notice has been given by virtue of which it will come to an end within fourteen months.

The competent landlord is the person entitled to give and receive notices and conduct negotiations under Part II.

(b) Mesne landlord

The activities of the competent landlord are binding on all mesne landlords holding interests between those of the competent landlord and his tenant. Where notices are given or agreements reached without the consent of the mesne landlord or landlords, they are entitled to compensation from the competent landlord for any consequential loss. Where the competent landlord seeks consent, that consent shall not be unreasonably withheld although it may be given subject to reasonable conditions including modification of the proposed notice or agreement or the payment of compensation. Questions as to whether consent has been unreasonably withheld will be determined by the courts.

(c) Superior landlord

Where the competent landlord is not the freeholder and his interest is a tenancy which will come or can be brought to an end within sixteen months and he gives the tenant notice to quit or receives a request from the tenant for a new tenancy, he is required to send a copy to his immediate landlord who, if he too is a tenant must send a copy to his immediate landlord also.

A superior landlord who becomes the competent landlord within two months of a notice to quit given under section 25 may give notice that he withdraws the notice previously given in which case it shall cease to have effect. He is not barred from giving a further notice under the Act.

7.3 TENANCIES EXCLUDED

There are specific exclusions so that the Act does not apply to:

- a tenancy of an agricultural holding;
- a tenancy created by a mining lease: defined as a lease for any mining purpose or purposes, while mining purposes include the sinking and searching for, winning, working, getting, making merchantable, smelting or otherwise converting or working for the purpose of any manufacture, carrying away and disposing of mines and materials in or under land and the erection of buildings and the execution of engineering and other works suitable for those purposes;
- a tenancy protected by the Rent Acts (or which would have been but for the fact that the rent reserved is less than two thirds of the rateable value) and entered into before 1 April 1990; on and after that date, a tenancy is not protected under the Rent Acts where the rent is £1000 or less per annum;
- a tenancy of premises licensed for the sale of intoxicating liquor for consumption on the premises was excluded until the law was changed by the Landlord and Tenant (Licensed Premises) Act 1990; any tenancy entered into after 11 July 1989 is protected: any tenant of a lease commencing before that date became entitled to protection from 11 July 1992;
- a tenancy granted by reason that the tenant was holder of an office, appointment or employment from the grantor which continues only so long as the tenant holds that position; where the tenancy was created after the commencement of the 1954 Act, the exemption will apply only where the tenancy was granted by an instrument in writing which expressed the purpose for which the tenancy was granted;

- a tenancy granted for a term certain not exceeding six months unless the tenancy contains provisions for renewing the term or extending it beyond six months or the tenant and his predecessor in the business have been in occupation for a period exceeding twelve months.

In addition certain specified landlords are exempt from the provisions of the Act by virtue of sections 57 and 60. Included are government departments, local authorities, statutory undertakers, development corporations and other public bodies.

A tenancy at will is excepted and in certain cases the provisions of sections 24 to 28 of the Act may be excluded by agreement between the parties (contracting out).

7.4 CONTRACTING OUT

The parties to a lease are not in general in a position to enter into any agreement which purports to preclude the tenant from making an application or a request for a new tenancy or provides for the termination or the surrender of the tenancy on making such an application or request or imposes any penalty on the tenant for so doing. Any such provision will be void although it will not invalidate the remainder of the agreement.

An exception is made so as to enable the landlord and tenant to enter into an agreement for the grant of a future tenancy of the holding or of the holding with other land on terms and from a date specified in the agreement. This exception is necessary to sanction removal of the tenant's right to claim a new tenancy, a right no longer available once the agreement has been entered into. The current tenancy will then no longer be a tenancy to which Part II applies although it will continue in force until the start of the new tenancy by virtue of these provisions.

Similarly, any agreement purporting to restrict the right to compensation where the tenant or the tenant and his predecessor in the business have occupied the premises for the purposes of a business carried on by the occupier or for those and other purposes during the whole of the five years immediately preceding the date on which the tenant is to quit the holding shall be void. The parties may reach agreement as to the amount of compensation once the right has accrued. Where the above provisions are not satisfied, the right to compensation may be modified or excluded.

These rules were found in practice to preclude lettings where although the tenant recognised the need for the landlord to obtain possession at some specified future date, he was not in a position to enter into an undertaking to give up the premises when required. The 1969 Act therefore introduced a subsection to enable the parties to make a joint application to the court to authorise an agreement excluding the provisions of sections 24 to 28 of the 1954 Act and, again on a joint application, to authorise an agreement for the surrender of the tenancy on such date or in such circumstances and on such terms, if any, as may be specified. The practical implications need to be considered, since it has been shown that the rental value with the tenant's rights to renew may be considerably greater than a tenancy without this provision.

It had been a common practice where tenants wish to assign to require them first to offer to surrender their lease to the landlord: on the landlord's refusal to accept a surrender or failure to take up the offer, the tenant was free to assign, subject to the landlord's consent but not to be withheld unreasonably. It has now been held that such a provision is void because it precludes the tenant from making an application or request under Part II or provides for the termination or surrender of the tenancy in that event contrary to section 38(1) of the Act (*Allnatt (London) Properties Ltd* v. *Newton* (1980)). The surrender must not take place in the circumstances described in section 24(2)(b) nor should it fetter his right to apply for a new tenancy (in which case any agreement would be void under section 38(1) of the 1954 Act). It has been held that an agreement contained in a letter from the landlord and countersigned by the tenant whereby the tenant agreed to vacate in consideration of release from outstanding rent should have been made by deed under section 52(1) of the Law of Property Act 1925 and the letter was not therefore an effective surrender.

7.5 CONTINUATION OF TENANCY

It is provided by section 24 that no tenancy to which Part II of the Act applies shall come to an end unless terminated according to the provisions of Part II. Even though the original lease is for a term certain, the tenancy will continue, if the tenant wishes it to do so and takes no further action, unless the landlord serves notice to terminate the tenancy under section 25 or the tenant makes a request for a new tenancy under section 26.

In either case there is an interim continuation of the tenancy for a period of three months beginning with the date on which the application is determined and finally disposed of by the court: were the landlord's notice to quit or the tenant's request for a new tenancy specified a later date, then that date will prevail.

Where two properties were held between the same parties on identical terms and used in one business, the Court of Appeal determined that a single originating application for a new tenancy was valid (*Curtis* v. *Galgary Investments Ltd* (1983)).

Where tenants wish to apply to the court for a new tenancy they must within two months of the date of the landlord's notice notify the landlord in writing that they are not willing to give up the tenancy. Unless agreement has been reached in the meantime, they must make application to court for a new tenancy not less than two nor more than four months after the landlord's notice or the tenants' request as the case may be.

An application to the court for a new tenancy has the effect of continuing the old tenancy for a period of three months beyond the date on which the application is finally disposed of but the period commences only from the date by which the proceedings on the application have been determined and any time for appealing has expired (section 64(1) and (2)). The effect is to leave a degree of uncertainty as to the expiry of the new lease granted by the court. This problem was overcome in *Chipperfield* v. *Shell UK Ltd* (1980) and *Warwick and Warwick (Philately) Ltd* v. *Shell UK Ltd* (1980) by an order that the new tenancies should end on a specified date.

It is essential that application to the court is made within the statutory period of four months. The case of *Stile Hall Properties Ltd* v. *Gooch* (1968) concerned a tenant's request which was not followed up in time. The court pointed out that the tenancy came to an end automatically at the date immediately before that specified by the tenant for the commencement of her new tenancy. As she had not commenced proceedings within the time allowed that put an end to the matter.

7.5.1 Interim rent

The landlord may apply to the court to determine an interim rent for the period during which the tenancy continues by virtue of section 24, payment at that rate to be made from the date on which the proceedings were commenced or the date specified in the landlord's notice or the tenant's request, whichever is the later.

The landlord's answer to the tenant's application for a new tenancy to the court may include an application for an interim rent to be determined. A landlord is not obliged to lodge an originating application in accordance with the county court rules (*Thomas* v. *Hammond–Lawrence* (1986)).

The court is required to determine a rent which it would be reasonable for the tenant to pay having regard to the rent payable under the terms of the tenancy but otherwise subject to subsections 1 and 2 of section 34 (the statutory disregards in assessing rent under a tenancy granted by order of the court) on the basis of a new tenancy from year to year of the whole of the property comprised in the tenancy.

In *Regis Property Co. Ltd* v. *Lewis & Peat Ltd* (1970), the Court took the view that it should have regard to the existing rent only where it provided some evidence of market value: in later cases this view was supplanted by the interpretation of Meggary J. who suggested that one purpose of the provisions for the assessment of interim rent was to enable a 'cushioning' effect where the market rent was considerably above the existing rent (*English Exporters (London) Ltd* v. *Eldonwall Ltd* (1973)). The courts have developed an approach which makes an allowance for the year to year nature of the hypothetical tenancy and a further allowance to soften the otherwise sharp increase in the rent payable.

The interpretation employed in particular cases ranges from the simple to the detailed. In *UDS Tailoring Ltd* v. *B.L. Holdings Ltd* (1981) where the existing rent was £5250, the new rent was assessed at £21 385.60 and reduced by 10% to £19 247.04 to reflect the year-to-year basis of the interim rent.

In another case, careful calculations by Finlay J. reduced the rent of £12 500 assessed for the new tenancy to £9200 for the amount payable as interim rent (*Janes (Gowns) Ltd* v. *Harlow Development Corporation* (1980)). In arriving at the latter figure, he deducted 10% from the original figure of £11 372 (which had been increased to £12 500 to take account of value increases since the end of the tenancy) to arrive at £10 235 and a further 10% to give £9212 (rounded to £9200) to reflect the 'tempering' effect of the statutory provision. (It is of some interest to note that the judge ordered an upwards or downwards rent review provision.)

The doubt that once existed as to the status of an interim rent application following withdrawal of an application for a new tenancy was resolved in *Kramer, Michael and Co.* v. *Airways Pension Fund* (1976). It was held that the interim rent application stands apart and survives withdrawal of the other application.

Where tenants wish to withdraw in a case within the jurisdiction of the High Court, they must obtain leave from the Court and the current tenancy will run for three months after leave has been given (*Covell Matthews and Partners* v. *French Wools Ltd* (1978)).

The interim rent should reflect any lack of repair for which the landlord is responsible. The court may fix an interim rent on the assumption that the premises are in a satisfactory state of repair but reduced so long as it takes for the necessary repairs to be completed (*Fawke* v. *Viscount Chelsea* (1979)).

In practice, it is usual to defer application for an interim rent pending the hearing of the landlord's application or the tenant's request so as to avoid the duplication of valuation evidence.

The valuation problems have been considered by the courts. In *English Exporters (London) Ltd* v. *Eldonwall Ltd* (1973) Megarry J. observed 'I would only add that the process of applying section 34 to a hypothetical yearly tenancy is one that, at least under present conditions, may often have an air of unreality about it and that would puzzle the most expert of valuers'.

But valuers are used to valuation constraints: one aspect of valuation concerns the adaptation of evidence derived from one transaction to the different circumstances of another.

In *Ratners (Jewellers) Ltd* v. *Lemnoll Ltd* (1980) both valuers first assessed the market rental value on the basis of a lease with five year reviews. The principles laid down in the *Eldonwall* case were applied with the result that, after hearing the evidence Dillon J. deducted 15% for the difference between a term of years and a yearly tenancy and 5% of the resultant sum for the stringency of the user clause. The arithmetic produced a figure of £19 582 which the judge reduced to £17 500 having regard to the existing rent (a reference to the requirements of section 24A of the 1954 Act and to the direction in subsection (3) that in determining an interim rent, the court shall have regard to the rent payable under the terms of the tenancy).

The judge remarked that where, as in this case '. . . two deductions fall to be made and the resultant figure has then to be tempered by having regard to the rent under the existing lease, mathematical precision in computing each deduction is hardly necessary'.

As noted above, once a new agreement is entered into the current tenancy is continued until the commencement of the new agreement by section 24 but is no longer subject to the provisions of Part II of the Act.

Section 24 does not affect a tenancy terminated by the tenant either by notice to quit or surrender or by forfeiture, though relief may be obtained from the court where the forfeiture is of a superior tenancy. In an endeavour to prevent avoidance or abuse of these provisions, there are restrictions on the issue of a notice to quit before the tenant had been in occupation under the tenancy for one month or, in the case of a surrender, the instrument was executed before, or was executed in pursuance of an agreement made before, the tenant had been in occupation under the tenancy for one month.

7.6 TERMINATION OF TENANCY

Apart from a tenant's request for a new tenancy, a tenancy may be terminated by:

- a notice to quit by the tenant;
- a notice to quit served by the landlord;
- surrender or forfeiture.

Where notice is served by the tenant, it must be of the proper length as required by the agreement. Where the tenancy is for a certain term, the tenant must serve notice not later than three months from the end of the term, otherwise the tenancy will continue in accordance with section 24.

The Act provides that notices should be sent by registered post while the Recorded Delivery Service Act 1962 modifies the requirement to enable such notices to be sent by the recorded delivery service. Where a notice was sent by ordinary post, it was held that tenants must prove service within the statutory period if they were not to lose their rights (*Chiswell* v. *Griffon Land and Estate Ltd* (1975)).

The Act is concerned mainly with the notice served by the landlord under section 25. There is a prescribed form which gives tenants notice of their rights and informs them whether the landlord is prepared to grant a new lease and if not, the reasons for wishing to obtain possession. A notice not in substantially the form prescribed would probably be bad, though the courts have upheld notices where the content of the notes forming part of the prescribed notice was inaccurate in that no alterations had been made to take account of amendments introduced by the 1969 Act. Should the tenant claim that they did not receive the notice the court would have to consider whether for the purposes of section 7 of the Interpretation Act 1978 there was sufficient evidence to prove the contrary (*Lex Services plc* v. *Johns* (1990)).

Before tenants can validly claim the protection of the 1954 Act, they must show that they are in possession for the purposes of the section – the correct test was first laid down in *Halberstam* v. *Tandelco Corpn NV* (1984) as to whether the thread of continuity of business user continued unbroken. This test was applied in *Aspinall Finance Ltd* v. *Viscount Chelsea* (1989) where the tenant surrendered a gaming licence in favour of other more suitable premises. As a result, the premises were empty when the tenant applied for a new tenancy although it claimed that it was its intention to reopen as a gaming club but was unable to gain the approval of the Gaming Board until a new lease had been negotiated. The tenant was held to remain in occupation.

In *Tegardine* v. *Brooks* (1977) surveyors served a notice to quit which was incomplete in that three of the notes were omitted as being irrelevant in the particular circumstances. The notice was held to be substantially to the same effect as the statutory form. In the course of the judgment, attention was drawn to a statement of the law made in *Bolton's (House Furnishers) Ltd* v. *Oppenheim* (1959):

> It is clear, I think . . . that this notice should be construed liberally and provided that it does give the real substance of the information required, then the mere omission of certain details, or the failure to embody in the notice the full provisions of the section of the Act referred to will not in fact invalidate the notice.

In *Pearson and another* v. *Aylo* (1990) a section 25 notice seeking possession of hotel premises was held to be invalid in not naming both the notice giver and his wife, who were co-owners. In the view of the court, the notice was capable of prejudicing a reasonable tenant.

In a case where the tenant was allowed into possession pending lease negotiations and who subsequently refused to provide a security deposit and was required to quit, the Court of Appeal granted an order for possession, reasoning that a statutory tenancy had not been created but that the tenant was a tenant at will pending resolution of the issue. Similar reasoning was applied where the plaintiffs intended to develop and therefore agreed with the tenants that the leases would be contracted out under the provisions of section 38(4), three short leases being arranged on this basis. At the expiry of the last term, extensions were negotiated but not made subject to a further court order. The landlords were able to obtain possession on the view of the court that the agreement between the parties was a tenancy at will.

According to the test laid down in *Carradine Properties Ltd* v. *Aslam* (1976) and adopted in *Germax Securities Ltd* v. *Spiegal* (1978) the questions to be put are 'Is the notice quite clear to a reasonable tenant reading it?' and 'Is it plain that they cannot be misled by it?'

The matter was considered once again in *Philipson-Stow* v. *Trevor Square Ltd* (1980). After a review of the authorities, Goulding J. concluded that a notice which, though inaccurately expressed, was not deceptive or misleading and which made clear which of the seven grounds the landlord was relying on was valid.

As a first step, tenants must notify the landlord in writing, within two months of the giving of the notice, whether or not, at the date of termination, they will be willing to give up possession. Secondly, if they wish to obtain an order for the grant of a new tenancy, they must apply to the court not less than two or more than four months after the landlord has given notice.

Failure to respect the timetable laid down will preclude any further action on the part of the tenant. Solicitors served a section 25 notice in which they were named as agents for the landlord, his name being omitted. On appeal it was held that the prescribed form required the name of the landlord to be shown (*Morrow* v. *Nadeem* (1986)). A 'harsh and regrettable' result was unavoidable in the case of *Smith and others* v. *Draper* (1990) where the tenant acted upon an invalid notice but when the landlords realised that the notice was invalid served a further notice to which the tenant replied that he was unwilling to give up the tenancy but failed to apply to the court within the statutory time limit for a new tenancy.

Where the court is closed by authority on the final day of the period of notice, it has been held that the tenant is entitled to give notice on the next working day – in this case the day after Easter Monday (*Hodgson* v. *Armstrong and another* (1966)).

It is important to serve any notice as provided by the Act: a section 25 notice must be served not more than 12 nor less than six months before the date it specifies for the termination of the current tenancy which cannot be earlier than the date on which it could have been terminated but for the Act. Furthermore, as the following selection of cases will show, it is important that the requirements are strictly observed. Two cases decided in 1983 illustrate the fine dividing line and surely emphasise the need to follow the procedure required by the Act to the letter, even where other negotiations in hand may render them unnecessary in due course.

In *Lewington* v. *Trustees for the Protection of Ancient Buildings* (1983) the landlords served a section 25 notice but the tenant had not responded with a counternotice mainly because negotiations were taking place for the purchase of the premises by the tenant, a price having been agreed and draft contract, conveyance and deposit being in the hands of the landlord's solicitors. Whilst the actions were held to be sufficient evidence of the tenant's intentions not to leave, the case was distinguished in *Mehmet* v. *Dawson* (1983) where the tenant had merely indicated his willingness to negotiate purchase of the premises with the landlord.

It was stated in *Robert Baxendale Ltd* v. *Davstone Holdings Ltd* (1982) that notices could be accepted out of time where the circumstances so warranted and lack of service was not entirely the fault of the tenant. An example of such circumstances occurred in *Ali* v. *Knight* (1984) where the tenant's originating application for a new tenancy was issued but not served in recognition of both parties acceptance that negotiations were continuing. When the landlord eventually applied to strike out proceedings for want of service, the Court of Appeal upheld the action of the deputy registrar in extending the time for service in the circumstances disclosed.

In *Hogg Bullimore and Co.* v. *Cooperative Insurance Society Ltd* (1984) it was held that, in computing whether the requirement had been complied with, only one of the dates should be excluded, not both. In respect of an application to the court for a new tenancy was held to be valid, having been made on the corresponding date two months after it was given (*Riley, E.J. Investments Ltd* v. *Eurostile Holdings Ltd* (1985)).

The requirements in respect of a section 25 notice served by the landlord but not received by the tenant were discussed in *Italia Holdings SA* v. *Bayada* (1985) where the notice was sent by recorded delivery to the last known place of abode of the tenant as required in section 23 of the Landlord and Tenant Act 1927.

A notice served by solicitors and signed by them as agents for the client (who was the controlling shareholder of the landlord company) was held to be invalid, even though he was authorised to act on behalf of the company (*Morrow* v. *Nadeem*, above).

In *Lex Services plc* v. *Johns* (1990) the tenant denied having received a section 25 notice sent by recorded delivery although the delivery book contained an illegible signature. The Court of Appeal held that the notice had been properly served, indicating that it would look for positive evidence of return in the case of a registered letter or where sent by registered letter or recorded delivery that there is no acknowledgement of it having been received by or brought to the attention of the addressee.

Negotiations between landlord and tenant were successfully concluded in *Stratton, R.J. Ltd* v. *Wallis Tomlin & Co. Ltd* (1985) and confirmed in a 'without prejudice' letter written by the landlord's agent who enclosed a copy of the landlord's open letter signifying agreement. The landlords went into receivership and the premises were marketed and conveyed within a very short time. It was held that the original parties had reached an agreement but it was not binding on the new landlord as it had not been registered as an estate contract. In any event, the agreement was within sections 28 and 62(2) of the 1954 Act as a result of which the tenants no longer had any rights under the Act.

Smith and others v. *Draper* (1990) produced a harsh result for the tenant who responded to a section 25 notice by a counternotice and an application to the court for a new tenancy before the landlord's solicitors served a further notice naming all three co-owners, only two owners having been named in the first notice; at the same time, the solicitors claimed that the first notice remained valid. The tenant served a further counternotice but did not follow up this time with an application to the court. Solicitors for the landlord informed the tenant that they had now abandoned the first notice and suggested that she withdraw the court proceedings. Although she did not do so, the landlord gained possession because the tenant had, in the view of the Court of Appeal, failed to protect her rights under the Act.

Where a renewal for three quarters was granted by consent and the landlord followed up with a section 25 notice, purporting to terminate the lease almost one year after the term agreed, it was held that the term did not continue under section 24 as the tenant was not in occupation for business purposes at the time. It also confirmed that even where tenants have received a section 25 notice but do not wish to remain, they can serve a three months' notice under section 28(2) to expire on any quarter day (*Long Acre Securities Ltd* v. *Electro Acoustic Industries Ltd* (1990)).

A landlord assigned the reversion to a wholly-owned subsidiary but did not inform the tenant. Shortly thereafter, a section 25 notice was served in the name of the original owner, terminating the tenancy whilst the tenant served a counternotice, seeking a new tenancy. The notice to quit was held to be invalid since it had not been given by the landlord (*Yamaha-Kemble Music (UK) Ltd* v. *ARC Properties Ltd* (1990)).

There are prescribed forms for the various notices set out in regulations under the 1954 Act and listed at the end of this chapter.

7.6.1 Treatment of break clauses

It is sometimes convenient to the parties or to one of the parties to be able to break the lease before it has run its full term. Such a provision in step with the rent review interval can give comfort to tenants who have a concern that they may not be able to afford the new rent; the landlord is protected where he grants a lease but wishes to reserve the right to re-enter in the event that redevelopment is possible or contemplated.

The break may be exercisable by either one or by both parties and the lease should make this clear and also when and how it may be initiated.

As far as the landlord is concerned, he is still required to serve a section 25 notice on the tenant evincing the appropriate ground for wishing to terminate the lease (which should, in addition, comply with any limitations imposed by the lease clause). The tenant therefore retains the full protection of the Act. By contrast, should the tenant decide to take advantage of the provision, he need do no more than serve a notice to quit in compliance with the provisions of the lease and in the form specified by section 27 of the Landlord and Tenant Act 1954. The courts may impose a break condition in the new lease as part of the order. Where there was some likelihood of obtaining planning permission but short of the proof required to satisfy section 30 (1)(f), a new lease for a term of 14 years was ordered subject to the right of the lessor to terminate the lease at an earlier date on giving not less than six months' notice. The court paid particular attention to the additional costs of financing in the event of the development being delayed (*National Car Parks Ltd* v. *The Paternoster Consortium Ltd* (1989)). This approach is safer than the result obtained in *Gregson* v. *Cyril Lord* (1962), considered later, despite its reference to the prospect of obtaining planning permission. The alternative is to grant a relatively short term in acknowledgement of the likelihood of development, rather than order a break clause to be inserted (*Becker* v. *Hill Street Properties Ltd* (1990)).

The report of *City Offices (Regent Street) Ltd* v. *Europa Acceptance Group plc* (1990) demonstrates that the validity of a section 25 notice is a matter for the scrutiny of the court even when originating from the provisions of a break clause. In another case where a break clause was contained in the original lease, the landlords proposed various concessions but insisted on retention of a break clause in the new lease, a requirement endorsed by Mr Justice Hoffman who entered judgement in the terms proposed by the landlord (*Leslie & Godwin Investments Ltd* v. *Prudential Assurance Co. Ltd* (1987)).

In drafting the break clause, it is important that it has regard to the other provisions of the lease; for example, it is commonly provided for maintenance to be carried out at intervals during the lease 'and in the final year of the term hereby granted'. If tenants are not to avoid their obligations, clear words must be used to provide for the effect of such provisions should the tenant exercise a right to terminate under the break clause (*Dickinson* v. *St Aubyn* (1994)).

7.7 GROUNDS FOR OPPOSITION TO NEW TENANCY

Notice to quit is often served simply as a necessary preliminary to renegotiating the terms of the lease and in particular the rent payable thereunder. The landlord is required to state as part of the notice whether he would oppose an application to the court for the grant of a new tenancy. Where tenants wish to gain a new tenancy, they must serve a counter-notice within the period laid down to protect their rights should agreement prove impossible. Failure to do so or, where that has been done, failure to preserve their rights by application to the court where no new terms have been agreed will almost certainly result in the tenants losing the protection of the act.

Where tenants are willing to give up possession they need take no action and will vacate in accordance with the notice.

Where the landlord wishes to obtain possession, he must state in the notice that he would oppose an application to the court for the grant of a new tenancy, stating the ground or grounds on which he relies. Section 30(1) of the Act sets out the seven grounds on which the landlord may oppose an application. He may quote as many as he believes to be applicable but it should be noted that if a strong case can be made under one or more of the first four grounds the landlord would not be liable to pay compensation to the tenant.

7.7.1   Ground (a) – state of repair

Where under the current tenancy the tenant has any obligations as respects the repair and maintenance of the holding, that the tenant ought not to be granted a new tenancy in view of the state of repair of the holding, being a state resulting from the tenant's failure to comply with the said obligations.

The court has discretion, the exercise of which will be influenced, no doubt, by the severity of the breach. In *Lyons* v. *Central Commercial Properties Ltd* (1958) it was held that serious breaches had occurred and had not been remedied and that relief ought not to be granted. The tenant's negotiations to transfer his interest to a third party were not relevant to the court's determination.

7.7.2   Ground (b) – persistent delay in paying rent

That the tenant ought not to be granted a new tenancy in view of persistent delay in paying rent which has become due.

There must be not only delay but persistent delay: the implication is that the delay must occur on more than one isolated occasion: the precise interpretation is left to the discretion of the court. The frequency and length of delay are relevant as are the additional management costs incurred by the landlord as a result of the delay (*Hopcutt* v. *Carver* (1969)).

Landlords opposed to the grant of a new tenancy originally cited grounds (a) and (b) but subsequently contended that the tenant had not occupied the premises, the restaurant having been occupied by a series of companies which had been incorporated and then wound up. The tenant took only a minor part in the running of the business although holding shares. It was held that occupation by the company was not equivalent to occupation by the tenant for the purposes of section 23(1) (*Norzari–Zadeh* v. *Pearl Assurance plc* (1987)).

In *Hurstfell Ltd* v. *Leicester Square Property Ltd* (1988) the tenant made numerous late payments, the periods of delay ranging from 4 to 19 weeks; on two occasions the landlord commenced proceedings for recovery. Following service of a section 25 notice the tenant paid rents promptly when due and the conclusion of the county court judge that further rent arrears would not occur was upheld by the Court of Appeal with the effect that the ground of appeal was not established. One route open to the court is to require that the tenant provide a surety or guarantor of his obligation (*Cairnplace Ltd* v. *CBL (Property Investment) Co. Ltd* (1983)).

In a further case, the Court of Appeal dismissed the tenant's appeal on the basis that he made no attempt to offer to pay future rent in advance or give security for payment.

In general, appeals are unlikely to be successful unless it can be shown that the lower court could be shown to have come to the wrong conclusion on the facts. It is clear from the decided cases that the delay should be substantial and persistent and that even regular late payments of a few weeks would be unlikely to establish the ground to the satisfaction of the court. It is interesting to note that a bad history may be redeemed by subsequent action.

7.7.3 Ground (c) – other substantial breaches

That tenants ought not to be granted a new tenancy in view of other substantial breaches by them of their obligations under the current tenancy, or for any other reason connected with the tenants' use or management of the holding.

Whether a breach is substantial is a matter of fact on the particular circumstances including any waiver by the landlord (but see *Norton* v. *Charles Deane Productions Ltd* (1970): a breach of covenant may be insufficient as a cause for obtaining forfeiture of a lease but may be adequate for refusing renewal of a tenancy).

The Court of Appeal held that in considering grounds a, b and c, it was entitled to consider the whole of tenants' conduct in relation to their obligations under the tenancy and need not limit itself to the grounds stated in the landlord's notice (*Eichner* v. *Midland Bank Executor and Trustee Co. Ltd* (1970)). This is a factual matter on which the court has a discretion.

The court has a discretion which was considered by Ormerod L.J. in *Lyons* v. *Central Commercial Properties Ltd* (1958) '. . . without attempting to define the precise limits of that discretion the judge, as I see it, may have regard to the conduct of the tenant in relation to his obligation'.

The reference in this ground to the tenant's use or management of the holding is a useful additional provision. In a case where a tenant was using premises in breach of an enforcement notice under the Town and Country Planning Act 1971 and where he intended to continue doing so if granted a new tenancy, it was held that the court could not condone the tenant's illegal conduct by making an order for a new tenancy which would create an illegal contract (*Turner and Bell* v. *Searles (Stanford-le-Hope) Ltd* (1977)). This was seen as one of the 'any other reasons' of the ground.

Where, despite a covenant not to use land for business purposes, the tenant did so use part of the land and then claimed a new tenancy which the landlord opposed under this ground, the breach of covenant was sufficient to deny the grant of a new tenancy, it being not possible to infer a waiver since the breach was a continuing one.

7.7.4 Ground (d) – provision of alternative accommodation

That the landlord has offered and is willing to provide or secure the provision of alternative accommodation for the tenant, that the terms on which the alternative accommodation is available are reasonable having regard to the terms of the current tenancy and to all other relevant circumstances and that the accommodation and the time at which it will be available are suitable for the tenants' requirements (including the requirement to preserve goodwill) having regard to the nature and class of his business and to the situation and extent of, and facilities afforded by, the holding. It may be that the offer on reasonable terms of part only of the premises presently occupied would be regarded in certain circumstances as suitable alternative accommodation.

The ground is otherwise unlikely to be much used except where the landlord wishes to redevelop or occupy the premises and is able to offer suitable premises nearby.

7.7.5 Ground (e) – uneconomic letting of part

Where the current tenancy was created by the subletting of part only of the property comprised in a superior tenancy and the landlord is the owner of an interest in reversion expectant on the termination of that superior tenancy, that the aggregate of the rents reasonably obtainable on separate lettings of the holding and the remainder of that property would be substantially less than the rent reasonably obtainable on a letting of that property as a whole, that on the termination of the current tenancy the landlord requires possession of the holding for the purpose of letting or otherwise disposing of the said property as a whole, and that in view thereof the tenant ought not to be granted a new tenancy.

This is essentially a financial or economic ground. The landlord needs to show that his return would be substantially improved by letting or disposing of the property as a whole. This is not a ground much used.

In *Greaves Organisation Ltd* v. *Stanhope Gate Property Co. Ltd* (1973) the landlords' claim for possession failed because they were unable to establish that the aggregate of the rents would be substantially less than if let as a whole.

7.7.6 Ground (f) – intention to demolish or reconstruct

That on the termination of the current tenancy the landlord intends to demolish or reconstruct the premises comprised in the holding or a substantial part of those premises or to carry out substantial work of construction on the holding or part thereof and that he could not reasonably do so without obtaining possession of the holding.

The effect of this ground is altered substantially by the provisions of a new section 31A introduced by the Law of Property Act 1969 which provides, in part:

1. 31A-(1) Where the landlord opposes an application under s 24(1) of this Act on the ground specified in para (f) of s 30(1) of this Act the court shall not hold that the landlord could not reasonably carry out the demolition, reconstruction or work of construction intended without obtaining possession of the holding if –

   - the tenant agrees to the inclusion in the terms of the new tenancy of terms giving the landlord access and other facilities for carrying out the work intended and, given that access and those facilities, the landlord could reasonably carry out the work without obtaining possession of the holding and without interfering to a substantial extent or for a substantial time with the use of the holding for the purposes of the business carried on by the tenant: or
   - the tenant is willing to accept a tenancy of an economically separable part of the holding and either para(a) of this section is satisfied with respect to that part or possession of the remainder of the holding would be reasonably sufficient to enable the landlord to carry out the intended work.

The section provides in subsection (2) a definition of 'economically separable part', no doubt thereby avoiding much discussion and argument as to what is and what is not to be regarded as separable under this section:

> (2) For the purposes of subs (1)(b) of this section a part of a holding shall be deemed to be an economically separable part if, and only if, the aggregate of the rents which, after the completion of the intended work, would be reasonably obtainable on separate lettings of that part and the remainder of the premises affected by or resulting from the work would not be substantially less than the rent which would then be reasonably obtainable on a letting of those premises as a whole.

The effect of section 31A on section 30 (1)(f) was considered in *Blackburn* v. *Hussan* (1988) where the landlord wished to make substantial alterations involving the subject property and two other shop premises, a passage and lavatories to make one open area, when for a minimum period of eight weeks it would not be possible to operate a business, Parker L.J. expressed obiter his support for the view that section 31A did not contemplate an agreement where the destruction of the subject matter of the original holding was involved. Where the court found that it was not economic to reinstate dilapidated premises at a cost of £28 000 where the cost of rebuilding was £45 000, it was held that the works could not be carried out without obtaining possession 'without interfering to a substantial extent or for a substantial time with the use of the holding for the purposes of the business carried on by the tenant' and that therefore section 31A(1)(a) was inapplicable (*Mularczyk* v. *Azralnove Investments Ltd* (1985)).

Where the landlords served a notice requiring possession 'to carry out substantial work on construction on the holding' amounting to infilling a pit created by the extraction of clay for brickmaking, it was held in *Botterill* v. *Bedfordshire County Council* (1984) that such work would alter the shape of the land but would not result in 'works of construction'.

In *Leathwoods Ltd* v. *Total Oil (Great Britain) Ltd* (1984) the landlords wished to gain possession to demolish and redevelop the site which the tenant claimed could be achieved under the lease which permitted the landlord to enter and carry out improvements, additions or alterations to the premises. The Court of Appeal upheld the view that legal as well as physical possession was necessary to enable the landlords to carry out the work which would amount to a derogation of grant if done under the provisions of the lease.

In *Aireps Ltd* v. *City of Bradford Metropolitan Council* (1985) the landlords exercised a right in the lease to take possession of the accommodation for reconstruction, providing temporary accommodation for the tenant, but later served a section 25 notice stating that a new tenancy would be opposed. The Court of Appeal held that the application was inadmissible since the landlords had obtained possession and the premises no longer existed.

In *Cerex Jewels Ltd* v. *Peachey Property Corporation plc* (1986) the lease reserved a power of entry but this would have enabled the landlord to carry out only part of the works and the question of whether the works could be carried out whilst the tenant remained in occupation was again raised. The tenant's appeal was allowed in this case.

Evidence that planning permission had been granted and information from the director of a building company and that the development was similar to a successful completed project, that he was interested in the project and had obtained the support of the bank in principle was sufficient to defeat the tenant's claim for a new tenancy (*Capocci* v. *Goble* (1987)).

'Intention' (in relation to the landlord's intention to demolish or reconstruct) was defined by Asquith L.J. as the need for the proposal to have 'moved out of the zone of contemplation – out of the sphere of the tentative, the provisional and the exploratory – into the valley of decision'. Intention must be established as at the date of the hearing of the application (*Betty's Cafes Ltd* v. *Phillips Furnishing Stores Ltd* (1958)).

Lord Justice Denning said that an 'intention' to carry out work 'connotes an ability to carry it into effect' (*Reohorn* v. *Barry Corporation* (1956)).

The more firm the landlord's proposal, the better the chance of proving an intention. Architects' plans, estimates, planning permission will all be persuasive although the main question is not whether the landlord has permission but rather the prospects of obtaining permission (*Gregson* v. *Cyril Lord* (1962)). The test there was said to be whether a reasonable person would believe, on the evidence, that there was a reasonable prospect of getting that permission or consent.

Intention should be firmly evidenced: for example in the case of a company by a board minute.

In *Bolton, H.L. Engineering Co. Ltd* v. *T.J. Graham and Sons Ltd* (1956) it was held that the intention of three directors was the intention of the company, even though there had been no board meeting. A similar conclusion was reached relating to the intention of a local authority (*Poppett's (Caterers) Ltd* v. *Maidenhead Borough Council* (1971)) but it is clearly safer to have a formal resolution.

The meaning of 'intention' was considered again in *Edwards* v. *Thompson* (1990) where the court held that a landlord of business premises could not successfully oppose her tenant's application for a new tenancy on the ground that he wished to redevelop the holding, if he could show that he had the means and ability to carry out the proposed development in its entirety. Here the proposals extended to other land in his ownership where she had not produced detailed costings or selected a developer. It was necessary in this case to look at the overall picture since one of the planning conditions required that the whole development, including an access road, was completed.

Reconstruction envisages substantial structural work (*Joel* v. *Swaddle* (1957) where a proposal to convert two single shop units to form part of an amusement arcade was held to be reconstruction) or rebuilding (*Cadle, Percy E. and Co. Ltd* v. *Jacmarch Properties Ltd* (1957) where a proposal to combine into a self-contained unit three floors previously held separately was considered not to be reconstruction). The ground failed in *Heath* v. *Drown* (1972) where the landlord had power by virtue of the current tenancy agreement to enter under a repair clause and carry out necessary work.

In *Barth* v. *Pritchard* (1989) the Court of Appeal provided clarification of the Act in expressing the opinion that the position as a whole had to be considered insofar as works not themselves works of construction were directly related or ancillary to works which were. It was held that works had to involve the structure of the building and in this particular case stated that even if they were wrong the works were not substantial. The saving grace of this restrictive interpretation from the landlord's point of view is that each case contains much factual information and must be considered on its merits.

An unusual case is noted here. A tenant company applied to the County Court for the outstanding terms of a new lease to be determined: the judge ordered a five-year lease with a break at three years based entirely on detailed evidence of the intention of the plaintiff company to redevelop the premises – yet earlier the landlords had not opposed the grant of a new tenancy or called in aid section 30(1)(f). The Court of Appeal dismissed the tenant company's appeal in what must be regarded as an unsatisfactory case on the facts disclosed (*Amika Motors Ltd* v. *Colebrook Holdings Ltd* (1981)).

The courts are not prepared to countenance an application which seeks to 'play the system' to gain further delays (*Smith, A.J.A. Transport Ltd* v. *British Railways Board* (1980)).

In *Edwards* v. *Thompson* (1990) the owner had obtained planning permission to develop a smithy occupied by a tenant together with a barn into one dwelling and develop other land with five dwellings subject to a condition that no part should be occupied until the development had been completed to include the access road. The smithy and barn were sold and the new owner served a section 25 notice stating that she would oppose the grant of a new tenancy under ground (f). She was able to show that a builder had been selected to carry out specified work and that finance was available. But because she could not show similar progress in developing the remainder of the site by the original owner, the Court of Appeal held that she was unable to show a firm and settled intention to develop the site, it being unlikely that a developer would be available for the remainder of the site at the termination date of the tenancy.

It is clear from decided cases that the position as a whole is to be considered to the extent that the works not being works of construction are directly related to or ancillary to works of construction. This approach was re-emphasised in *Romulus Trading Co. Ltd* v. *Henry Smith's Charity Trustees* (1990) where the Court of Appeal expressed the view that in order for works to qualify as reconstruction they must be shown as works of rebuilding involving a substantial interference with the structure of the building.

7.7.7 Ground (g) – landlord's intention to occupy

Subject as hereinafter provided, that on the termination of the current tenancy the landlord intends to occupy the holding for the purposes, or partly for the purposes, of a business to be carried on by him therein, or as his residence.

The case of *Gregson* v. *Cyril Lord* (1962) laid down two tests to be proved by the landlord in any attempt to obtain possession under this ground. The first was a genuine intention to occupy and the second the ability within their own powers of being able to achieve such occupation.

In *Westminster City Council* v. *British Waterways Board* (1985) the landlord sought possession from the tenants, the council which used the premises as a cleansing depot. The House of Lords upheld the Court of Appeal in holding that the council was not entitled to maintain possession in the light of the landlord's requirements by refusing planning permission to the landlord's proposals.

The evidence of the landlord's intention to operate a restaurant, of the existence of planning permission and the availability of finance enabled the company to satisfy the ground in *Chez Gerard Ltd* v. *Greene Ltd* (1983).

A firm and settled intention was also sufficient to satisfy the court in *Cunliffe* v. *Goodman* (1950) where the landlords produced minutes of board meetings, quotations for equipment required and an affidavit from the company's property director.

There is an important limiting proviso to the effect that the landlord shall not be entitled to oppose an application on this ground if the interest of the landlord was purchased or created after the beginning of the period of five years and ending with the termination of the current tenancy where at all times the tenancy has been one subject to Part II of the Act.

In *Cox* v. *Binfield* (1989) the only question to be decided where the landlord required the accommodation partly for her own business and partly for residential purposes was whether there was a genuine intention firm and settled and capable of being carried out in the reasonable future in the circumstances then prevailing. The Court of Appeal concluded that although the plans were ill thought out and might well fail, it could not be said that they were unrealistic to the point of not being genuine. Another case where the genuine intention of the landlord to run a business and the absence of corroborative evidence of that intention were submitted to be insufficient was not upheld by the Court of Appeal which dismissed the tenant's appeal. (*Mirz and another* v. *Nicola* (1990).)

It is clear from *Jones* v. *Jenkins* (1986) that a landlord's notice to obtain possession where the landlord would take possession only whilst the premises were converted to flats which would then be relet, would not be sufficient to satisfy this ground.

The ground cannot be invoked where the landlord has acquired his interest within the five-year period expiring on the date the notice is stated to take effect, except when the current landlord granted the term. In *Morar* v. *Chauhan* (1985) the landlord had created a family trust of the property but it was held by the Court of Appeal that this did not prevent him from opposing the grant of a new tenancy.

The landlord must have a firm intention to occupy and evidence of his proposals for alterations would be helpful in support of his intentions under this ground.

The intention to occupy must be more than a device to obtain possession and sell after occupying for a short time (*Willis* v. *Association of Universities of the British Commonwealth* (1965)) and would be strengthened by an undertaking to the court on the part of the landlord to occupy in the event of obtaining possession (*Espresso Coffee Machine Co. Ltd* v. *Guardian Assurance Co. Ltd* (1959)).

The complicated series of transactions disclosed in *Wates Estate Agency Services Ltd* v. *Bartleys Ltd* (1989) boils down to an avoidance of the five-year rule where the interest was gained by acquiring the shares of the company holding the headlease and which had granted the sublease to the occupier now being displaced.

In *Method Developments Ltd* v. *Jones* (1971) the landlords succeeded under this ground even though they did not intend to occupy the whole of the premises immediately and would leave a part unused. But where it was proposed to obtain possession of a collection of buildings and a yard and to incorporate the cleared land with other land in order to build a filling station on the combined site, the landlords did not succeed in their claim (*Nursey* v. *P. Currie (Dartford) Ltd* (1959)).

In a more recent case, the landlord sought to obtain possession to enable him to erect a building on what was then a car park operated by the tenant in conjunction with his business on adjoining land. The court distinguished Nursey on the grounds that the landlord merely intended to place a building on the site and not to create a wider scheme and dismissed the tenant's appeal (*Cam Gears Ltd* v. *Cunningham* (1981)). The landlord may occupy by a manager or an agent (*Cafeteria (Keighley) Ltd* v. *Harrison* (1956)).

In *Thornton, J.W. Ltd* v. *Blacks Leisure Group plc* (1986) the tenants sought to resist the landlord's application for possession. They occupied small premises which adjoined premises in the landlord's possession and which the landlord wished to merge by removing partitions. They claimed that the premises would lose their identity if the proposed work was carried out. The court dismissed the tenants' contention.

Two solicitors who were beneficiaries of a trust together with a third person were unable to satisfy the ground to enable them to obtain the premises for occupation by the firm of which they were partners because their occupation would be by virtue of the lease and not as beneficiaries (*Meyer* v. *Riddick* (1989)).

In a case where the council landlord claimed possession as it wished to have control of the bowling operation but to run it through a management committee, the tenant lost the appeal. It was held that the business was to be carried on by the landlord, the council, as it retained the right to make policy decisions.

Under certain conditions, the landlord may obtain possession to enable a company which they control to occupy the premises (see section 7.10.1).

7.8   ORDER FOR GRANT OF A NEW TENANCY

If tenants are able to surmount the obstacles placed in their way by the reasons given for termination of a tenancy in a landlord's notice to quit or in reply to their request for a new tenancy, they are entitled to an order of the court for the grant of a tenancy for a term not exceeding fourteen years.

The court may, if it thinks fit, include a provision for varying the rent during the term of the new tenancy. Where the current tenancy includes rights enjoyed by the tenant in connection with the holding, such rights are to be included except as otherwise agreed or in default of agreement, determined by the court. This provision enables the landlord to make a case for the exclusion of rights previously enjoyed in connection with the holding under the current tenancy and removes the mandatory nature of the original subsection (3) of section 32. Where the tenant is willing to accept a tenancy of part of the holding, the order made by the court shall be an order for the grant of a new tenancy of that part only.

Where the court makes an order, the terms of the tenancy shall be such as may be agreed between the landlord and the tenant or, in default of agreement, determined by the court. The court is required to have regard to the terms of the current tenancy and to all relevant circumstances. The majority of disputes are confined to the rent payable, although there is often disagreement about the term to be granted, the frequency of rent reviews and the other terms to be included. Clearly, the terms agreed between the parties or fixed by the court will affect the rental value which should not therefore be determined until all the other terms are known.

The function of the court is to complete the agreement on terms where the parties have been unable to agree on all the terms. It is helpful to both parties that the terms already agreed will not be upset by the court and leaves them in greater control and with lower legal costs.

Where the rent is determined by the court it is to be the rent at which the holding might reasonably be expected to be let in the open market by a willing lessor, disregarding:

- any effect on rent of the fact that the tenant or his predecessors in title have been in occupation of the holding;

- any goodwill attached to the holding by reason of the carrying on thereat of the business of the tenant (whether by the tenant or by a predecessor of his in that business;
- any effect on rent of an improvement carried out by a person who at that time was the tenant but only if it was carried out otherwise than in pursuance of an obligation to his immediate landlord and either it was carried out during the current tenancy or the following conditions are satisfied:
- that it was completed not more than 21 years before the application for a new tenancy was made and
- that the holding or any part of it affected by the improvement has at all times since the completion of the improvement been comprised in tenancies which include premises which are occupied by the tenant and are so occupied for the purposes of a business carried on by him or for those and other purposes and
- that at the termination of each of those tenancies the tenant did not quit;
- in the case of a holding comprising licensed premises, any addition to its value attributable to the licence where it appears to the court that the benefit of the licence belongs to the tenant.

Trading accounts are admissible in evidence as to what a new tenant, an outsider, would pay although the accounts should not be used for the purpose of assessing what the present tenant could pay (*Harewood Hotels Ltd* v. *Harris* (1957)).

An important principle was established in *O'May and others* v. *City of London Real Property Co. Ltd* (1982). The defendant landlord proposed to amend the terms of an earlier lease on granting a new term of three years, so as to make the tenant liable for maintenance of the building and equipment and for depreciation of plant and equipment and suggested a reduction of the rent of 50 pence per square foot per annum to take account of this shift of burden. The attraction from the landlord company's point of view was that it would then have had a 'clear lease' which on the evidence would increase the capital value by between one and two million pounds by removing the speculative element in fluctuating costs for which the landlord was responsible. The valuers for landlord and tenant were able to agree that the reduction of 50 pence represented the additional liability cast on the tenants. Shaw L.J. opined that the landlord's revised terms introduced a radical change in the balance of rights and responsibilities, of advantage and detriment, of security and risk. Brightman L.J. agreed that this case had to be considered under section

35 and not section 34 because the substantial issue was not the amount or calculation of the rent but the incidence of certain unusual financial burdens which would have the effect of controlling the rent. He went on to remark that a short-term tenant is not adequately compensated by a small reduction in rent for their assumption of the financial risks implicit in the maintenance of the structure of an office block. Those risks should properly be borne by the owner of the inheritance: such risks are indeterminate in amount and could prove to be wholly out of proportion to the very limited interest held by a short-term tenant. Buckley L.J. thought that the new tenancy should, *mutatis mutandis*, be on similar terms to the existing tenancy. Any departure from those terms would require explanation, that is to say justification (*Cardshops Ltd* v. *Davies* (1971) where four tests were propounded by Goulding J. and adopted in the *O'May* case although there they led to an opposite conclusion). The reduction in the rental which was said to compensate the tenants for assuming those risks was far from being an indemnity against them. The tenants might find themselves saddled with very heavy capital expenditure for which the reduction in rent would by no means compensate them, except possibly in the very long run . . . which would be no comfort to a tenant for a relatively short period.

The four tests laid down by Goulding J. were:

1. Has the party seeking an alteration in the terms given a reason?
2. Is the adjustment of rent proposed adequate to compensate the party opposing the alteration?
3. Will the proposal adversely affect the security of the tenants?
4. If the answer to the first two questions is 'yes' and to the third 'no', does the proposed alteration appear to be fair and reasonable as between the parties?

A restriction on the use to which premises may be put may result in the determination of a reduced rent. In *Clements, Charles (London) Ltd* v. *Rank City Wall Ltd* (1978) the landlords sought to relax the covenant restricting use of the premises to the business of a retail cutler by adding to the expression 'without the landlord's consent in writing' a rider 'such consent not to be unreasonably withheld'. The tenants objected to the relaxation which would have had the effect of increasing the rental value. The tenants' objection was upheld.

In *Cairnplace Ltd* v. *CBL (Property Investment) Co. Ltd* (1983) it was held by the Court of Appeal that the judge was within his powers, when making an order for a new tenancy, to require as one of the conditions the provision of two sureties to guarantee the rent. A further condition requiring that the tenant should pay the legal costs in preparing the new lease was not upheld.

In *Chelsea Building Society* v. *R & A Millett (Shops) Ltd* (1993) an unusual situation occurred where the tenant required a renewal for a short term (under one year) whereas the landlords stipulated a 14-year period, giving evidence that such a short letting would affect the capital value of their long leasehold interest. Whilst accepting the landlords' views the court granted the short lease required by the tenant, reflecting the guidance in *O'May*.

Following the grant of a new tenancy on terms set by an assistant recorder, the landlord in *Khalique* v. *Law Land plc* (1989) applied for leave to appeal out of time, wishing to introduce new evidence of two comparables, one of which the assistant recorder had already excluded. Leave was refused. Where, in *Teltscher Bros Ltd* v. *London & India Dock Investments Ltd* (1989), the names of the plaintiff and defendant were transposed and service of the summons was refused on the ground that it was defective, the court held that the mistake was both genuine and not misleading and could be corrected by the court.

The question of whether the court should stipulate a break clause in an order for a new lease was considered in *National Car Parks Ltd* v. *The Paternoster Consortium Ltd* (1989) where the landlord had opposed the grant of a new lease. It was considered that there was a strong possibility of redevelopment taking place and that a 14-year term with a break clause enabling the lessor to terminate on giving not less than six months notice was appropriate.

A substantial difference between the new rent of £106 000 and a view as to the interim rent of £40 000 resulted from a difficult set of circumstances, where both landlord and tenant were initially under a misapprehension as to the true rental value. The judge made no order with regard to the interim rent as a result of certain undertakings by the tenant but in fixing a ten-year term he provided for an option on the part of the tenant to break on giving six months' notice, exercisable within one month of the commencement of the tenancy (*Charles Follett Ltd* v. *Cabtell Investment Co. Ltd* (1986)).

Inclusion in an order for a new tenancy of an upwards or downwards provision for the determination of rent on review is of particular interest in a falling market. Such a provision was included in the order made in *Amargee* v. *Barrowfen Properties Ltd* (1993).

7.9 COMPENSATION ON TERMINATION OF TENANCY

On leaving the premises tenants may be entitled to compensation for disturbance and also for any improvements carried out by them. They may also be able to claim compensation when subsequent events show that the court was induced to refuse an order for the grant of a new tenancy by misrepresentation or the concealment of material facts.

7.9.1 Compensation for disturbance

Where the only grounds specified in the landlord's notice relates to one or more of those contained in (e) (f) and (g) (section 7.7), the tenant is entitled to compensation. Grounds (a), (b), (c) and (d) do not attract compensation. Where more than one ground is stated and one of those grounds is from the latter group, the tenant's entitlement to compensation will turn on whether the court finds that it is precluded from making an order by reason only of the grounds set out in (e) to (g). The amount of compensation payable to the tenant is based on the rateable value of the premises on one of two bases at the option of the tenant where the lease was entered into before 1 April 1990 and notice to terminate is served by landlord or tenant after 1 April 1990 but before 1 April 2000, depending upon which rateable value is selected:

1. at the rate of one times the 1990 rateable value except where the tenant has been in occupation for 14 years or more when it will be twice the 1990 rateable value, or
2. at the rate of four times the 1973 rateable value except where the tenant has been in occupation for 14 years or more when it will be eight times the 1973 rateable value (Landlord and Tenant Act 1954 (Appropriate Multiplier) Order 1990 (SI 363/1990)).

Until 1969, tenants' application to court was a prerequisite of their entitlement to compensation. As a result of amendments introduced at that time, they are now entitled to compensation under grounds (e), (f) and (g) where no application is made or, having been made, is withdrawn. Should tenants wish to resist other grounds not entitling them to compensation they must still apply to the court.

Compensation payable to the tenant under section 37 of the Act is not chargeable to capital gains tax.

7.9.2 Compensation for misrepresentation

Where the court refuses an order for the grant of a new tenancy and it is subsequently shown that the court was induced to refuse the order by misrepresentation or by the concealment of material facts, the court may order the landlord to pay a sufficient sum as compensation for damage or loss sustained by the tenant.

There is no formula and no limit: the provision was undoubtedly intended to discourage irresponsible action by enabling the possibility of substantial damages being awarded. The provision makes it clear that the landlord involved in the application is responsible for payment and the tenant is eligible to make the claim: the burden and benefits do not pass to subsequent parties.

7.9.3 Compensation for improvements

The Landlord and Tenant Act 1927 survives, subject to certain amendments made by Part II of the 1954 Act, to enable the tenant to claim compensation for improvements in certain circumstances. Some differences of treatment apply where the contract was made before 10 December 1953: only the provisions relating to contracts entered into on or after the date are discussed here.

Where tenants propose to make an improvement they are required to serve on their landlord notice of their intention together with a specification and plan showing the proposed improvement and the part of the existing premises affected. The landlord may object to the proposals within three months after service of the notice whereupon the tenant may apply to the court for a certificate that the improvement is a proper improvement and the court must be satisfied that the improvement will add to the letting value of the premises at the termination of the tenancy. It must also be satisfied that the improvement is reasonable and suitable in character and will not diminish the value of any other property belonging to the same landlord or to any superior landlord from whom the immediate landlord holds directly or indirectly.

The court is empowered to make such modifications as it thinks fit or to impose such other conditions as it may think reasonable. The court shall not issue a certificate where the landlord shows that he has offered to carry out the work himself in consideration of a reasonable increase in rent or of such rent as the court may determine, unless it is shown subsequently that the landlord has failed to carry out the work. It is clear from *Historic Houses Hotels Ltd* v. *Cadogan Estates* (1993) that in the case of a tenant's application to court under section 3 of the 1927 Act for a certificate that the improvements were 'proper improvements', the works must not commence before the approval is received. Without such a certificate the tenant would be unable to claim compensation at the end of the tenancy.

Tenants are not entitled to claim compensation unless they have served notice of the proposal and obtained the agreement of their landlord or, in default thereof, of the court. The tenant may require the landlord to furnish him with a certificate to the effect that the improvement has been duly executed on payment by the tenant of any reasonable costs incurred by the landlord: where the landlord refuses or fails within one month after the tenant's request to furnish a certificate, the tenant may apply to the court. Possession of a certificate may avoid argument at a later stage when the tenant seeks to make a claim for compensation. Where a tenant carries out improvements after 1 October 1954 under a statutory obligation he may make a claim for compensation at the end of their tenancy: approval of the landlord is of course inappropriate although he must be given proper notice and the tenant may obtain at his own expense a certificate of execution. It is not clear whether the landlord may offer to carry out the work himself. Work carried out by the tenant in pursuance of a contractual obligation is outside the scope of the Act and does not qualify for compensation.

Having laid a proper basis for compensation for improvements, the tenant must observe the various time limits laid down for making a valid claim on quitting the premises. Where the tenancy expires by effluxion of time the tenant must serve a notice of claim not earlier than six nor later than three months before the end of the tenancy.

A claim where the tenancy is terminated by a notice to quit given by the landlord or the tenant must be made within three months from the date on which notice is given. Where the tenancy is terminated by a tenant's request for a new tenancy any claim must be made within three months from the date on which the landlord opposed the request or, where the request is not opposed, within three months from the latest date on which he could have opposed it. Where a tenancy is terminated by forfeiture or re-entry, the claim must be made within three months from the effective date of the court order for possession. Where the tenancy is terminated by re-entry without a court order, a similar time limit runs from the date of re-entry.

The amount of compensation for improvements may not exceed:

- the net addition to the value of the holding as a whole which may be determined to be the direct result of the improvement or
- the reasonable cost of carrying out the improvement at the termination of the tenancy, subject to deduction of an amount equal to the cost (if

any) of putting the works into a reasonable state of repair, except where such cost is part of the tenant's liability under his lease.

Any proposals by the landlord to demolish or alter the premises or to change their use are likely to have an adverse effect on the tenant's claim. Where compensation has been reduced or refused by the court following evidence of such proposals, the court may authorise a further application for compensation where effect is not given to the intention within a time fixed by the court.

7.10 MISCELLANEOUS MATTERS

The 1954 Act created unexpected problems for a landlord wishing to obtain possession for occupation by a business carried on by a company which he controls and for tenants in partnership. The 1969 Act introduced amendments designed to overcome the problems.

7.10.1 Companies

The provision (section 30(3)) enables the landlord to oppose a request for the grant of a new tenancy of premises on the ground that the company in which he has a controlling interest intends to occupy the premises for the purpose of its business.

For this purpose he has a controlling interest if either he is a member of the company and able to appoint or remove the directors or the majority of them or he holds more than one half of the company's equity share capital in his own right.

7.10.2 Partnerships

A case before the Court of Appeal in 1968 brought to light a problem where one of the joint tenants does not join in a tenant's request for a new tenancy. It was held that an application from one tenant was not valid since he was not the tenant for the purposes of section 24(1).

As a result, a new section (41A) was introduced to deal with the situation where a tenancy of business premises is held jointly by two or more persons all of whom at some time during the existence of the tenancy carried on a business (not necessarily the same business) but which is now carried on by one or some only of the joint tenant or tenants, the remaining tenant or tenants having no business occupation of any part of the property.

Form number	Purpose for which to be used
1(13)*	A notice under section 25 of the Act, being a notice terminating a tenancy to which Part II of the Act applies which does not contain a certificate given under the provisions of section 57, 58, 60, 60A or 60B of the Act.
2(14)*	A notice under section 25 of the Act, being a notice terminating a tenancy to which Part II of the Act applies, which contains a copy of a certificate given under section 57 of the Act (whereby the Minister or Board in charge of any Government department certifies that the use or occupation of the property or part of it shall be changed by a specified date) where the date of termination of the tenancy specified in the notice is not earlier than the date specified in the certificate.
3(15)*	A notice under section 25 of the Act, being a notice terminating a tenancy to which Part II of the Act applies, which contains a copy of a certificate given under section 57 of the Act (whereby the Minister or Board in charge of any Government department certifies that the use or occupation of the property or part of it shall be changed by a specified date) where the date of termination of the tenancy is earlier than the date specified in the certificate.
4	A notice under section 25 of the Act, being a notice terminating a tenancy to which Part II of the Act applies, which contains a copy of a certificate given under section 48 of the Act (whereby the Minister or Board in charge of any Government department certifies that for reasons of national security it is necessary that the use or occupation of the property should be discontinued or changed).
5(16)*	A notice under section 25 of the Act, being a notice terminating a tenancy to which Part II of the Act applies, which contains a certificate under section 60 of the Act (whereby the Secretary of State certifies that it is necessary or expedient for achieving the purpose mentioned in section 2(1) of the Local Employment Act 1972 that the use or occupation of the property should be changed).
6(17)*	A notice under section 25 of the Act, being a notice terminating a tenancy to which Part II of the Act applies, which contains a copy of a certificate given under section 60A (Welsh Development Agency premises) of the Act (whereby the Secretary of State certifies that it is necessary or expedient for the purpose of providing employment appropriate to the needs of the area in which the premises are situated, that the use or occupation of the property should be changed).
7(18)*	A notice under section 25 of the Act, being a notice terminating a tenancy to which Part II of the Act applies, which contains a copy of a certificate given under section 60B (Development Board for Rural Wales premises) of the Act (whereby the Secretary of State certifies that it is necessary or expedient for the purpose of providing employment appropriate to the needs of the area in which the premises are situated, that the use or occupation of the property should be changed).
8	A notice under section 26 of the Act, being a tenant's request for a new tenancy of premises to which Part II of the Act applies.
9	A notice under section 40(1) of the Act, being a notice requiring a tenant of business premises to give information as to his occupation of the premises and as to any sub-tenancies.
10	A notice served under section 40(2) of the Act on a landlord of business premises, being a notice requiring that landlord to give information about his interest in the premises.
11	A notice served under section 40(2) of the Act on a mortgagee in possession of business premises, being a notice requiring that mortgagee to give information about his mortgagor's interest in the premises.
12	A notice under section 44 of and paragraph 6 of Schedule 6 to the Act, being a notice withdrawing a previous notice given under section 25 of the Act to terminate a tenancy to which Part II of the Act applies.

Where a form number is shown in brackets the form with that number must be used if –
(a) no previous notice terminating the tenancy has been given under section 2i of the Act and
(b) the tenancy is the tenancy of a house (as defined for the purposes of part I of the Leasehold Reform Act 1967, and
(c) the tenancy is a long tenancy at a low rent (within the meaning of that Act of 1967), and
(d) the tenancy is not a company or other artificial person.

Figure 7.1 Notices revised by the Landlord and Tenant Act 1954, Part II (Notices) Regulations 1989 (SI No. 1548/1989).

The remaining business tenant or tenants are enabled to serve a valid tenant's request for a new tenancy (under section 26) or a notice to quit (under section 27) provided that it sets out the facts of the changed circumstances. The business tenant will be able to obtain a new tenancy either alone or jointly with other partners while the court may require guarantors or sureties as a condition of a new tenancy where appropriate. The business tenants will also be able to obtain statutory compensation and to exercise all other rights under the Act.

7.10.3   Notices under the Landlord and Tenant Act 1954

Prescribed notices under the Landlord and Tenant Act 1954 have been revised by the Landlord and Tenant Act 1954, Part II (Notices) Regulations 1989 (SI No. 1548/1989). A list of such notices is set out in Figure 7.1.

FURTHER READING

Fox-Andrews, J. (1987) *Business Tenancies*, 4th edn, Estates Gazette, London.
Haley, M. (1990) *Commercial Property*, Croner Publications Ltd.
Philpott, G. and Hicks, G. (1994) *Managing Business Tenants*, 1st edn, Estates Gazette, London.

Business tenancies: rent review and third party proceedings

8

Most modern leases contain provisions to ensure that the rent payable under the lease is reviewed at regular and frequent intervals, the objective being to maintain the purchasing power of the income. Where property investments have been purchased on the basis of low initial yields, a rising rent is essential in achieving anticipated returns. As the basis of the rent review is found in the lease, it is important that the provisions of the lease are known and interpreted properly and in sufficient time to ensure their operation to the benefit of the landlord. Similarly, tenants will wish to limit the increased rents payable under the lease and they or their advisors will look for interpretations of the provisions to assist that search.

The main areas of contention between the parties are considered and the outcome of a representative selection of cases included. There follows a brief account of dispute procedures available where the parties or their advisors fail to reach a satisfactory conclusion by negotiation.

8.1 INTRODUCTION

Provision for a review of the rent payable during the currency of a lease is now a common feature of leases. This is a development of the last thirty or so years which recognises the eroding effect of inflation on purchasing power and the need to protect future payments from its worst effects.

Occupation leases for a long term – typically 15, 21 or 25 years – offer security to both landlord and tenant but the onset of significant levels of inflation made landlords reluctant to grant long terms without the opportunity to adjust the rent to reflect changes in rental values (due to inflation or other changes in the value of the property and the immediate area). Legal advisers therefore began to incorporate rent review clauses in leases but without always achieving the intended effect. The continuing high level of litigation is testimony to the loose drafting of many review clauses, to the continuing ravages of inflation and, more recently, significant reductions in rental values from rents agreed only a few years ago by expansive owners and confident tenants.

It has been common for the landlord to insist on 'upward only' reviews which are hardly reasonable taking account of the uncertainty of the future and which may involve the tenant in onerous financial obligations.

Continuing trading problems being experienced by business firms may prepare the ground for a more general acceptance of upward or downward rent reviews although determined efforts are needed on the part of tenants if any significant shift is to be achieved.

The parties are required to agree the new rent payable in accordance with the provisions of the particular rent review clause. Such clauses have become increasingly sophisticated in attempts to exclude factors likely to have an adverse effect on the rental value. They may be assisted in their interpretation of the effect of the covenant by a number of recent judgements. The favoured provision is that of setting out the terms of a hypothetical lease although the courts have not always interpreted the provisions in the way intended. There is now a wealth of case law on the interpretation of the effect of a particular covenant which will be explored in this chapter. It is important to bear in mind, however, that the covenant must be read in relation to the other terms appearing in the particular lease in which it appears. Rent reviews of commercial premises have no statutory controls. The property manager may commission rental valuations as agent of either landlord or tenant, or for third party proceedings from a valuer acting as expert witness, or for presentation to an arbitrator or independent surveyor.

8.2 THE REVIEW CLAUSE

There is a wide variety of review clauses in use and, except where a review is included as one of the terms in any new lease granted by order of the court, the parties are free to agree whatever form appears appropriate although care must be taken to follow the requirements laid down in the lease.

8.2.1 Types of review clause

In practice, three basic forms have emerged. Many leases, particularly earlier ones, use as a basis the provisions of section 34 of the Landlord and Tenant Act 1954. These rather theoretical provisions were designed to guide the courts on tenancy renewals and are not entirely appropriate to the subject of reviews: not surprisingly, the rents resulting have often disappointed one or other of the parties when the precise meaning has been tested before the courts.

Leases granted by landlords experienced in property matters have not always avoided all the pitfalls, but in general have provided clearer and more specific directions for ascertaining the review rent. Indeed, there has been a tendency for long and elaborate clauses setting up a fictitious or hypothetical background against which the rent is to be adjudged.

The Royal Institution of Chartered Surveyors has combined with the Law Society to produce a model rent review clause with alternatives to suit particular circumstances whilst the Incorporated Society of Auctioneers and Valuers has made its own recommendations. If adopted on a large scale, many of the current problems would disappear eventually, although given the lifespan of many leases, it is inevitable that unsatisfactorily drafted review clauses will continue to present problems of interpretation for some time to come. There is no indication that this or any other attempt at producing a standard form, albeit with alternatives, is likely to attract wide support, given the unique nature of each property, the interaction of the review clause with other parts of the lease and the increasing complexity of the lease form, typically extending to 50 or more pages.

8.2.2 Form of notice

The service of correctly worded notices continues to cause problems as does the question of whether, in any particular case, time is of the essence. A letter intended to be a rent review notice in which a rent was specified was served by the landlords' solicitors in *Durham City Estates Ltd* v. *Felicetti* (1990). The following day a further letter was sent, specifying a rent in figures of one hundred pounds more than that in the original notice, although the words remained the same. In this case the Court of Appeal upheld the notice as a valid one and held that the proper test was whether the notice specified the amount of rent with sufficient clarity. In *Commission for the New Towns* v. *R. Levy & Co. Ltd* (1990) the trigger notice was held by the Court of Appeal to be defective since it did not indicate a rent whereas the rent review clause required that the notice 'shall' specify the proposed rent, a word regarded as mandatory, not merely directory. *Dean and Chapter of Chichester Cathedral* v. *Lennards Ltd* (1977) was distinguished because the rent review clause in that case admitted of a valid notice without the specification of a rent. The nature and form of the tenant's reply to the landlord's notice can be important. In *Museprime Properties Ltd* v. *Adhill Properties Ltd* (1990) it was held that only a written notice could be served and that telephone calls by the tenant were not valid as a response. In reply to the landlord's notice in *Patel* v. *Earlspring Properties Ltd* (1991) the tenant replied to the effect that the rent was beyond his means but did not specify a rent as required by the lease to be served within 28 days of the landlord's review notice. The Court of Appeal held that the requirement on the tenant to specify a rent was not mandatory and that it was still possible to refer the matter to arbitration.

In a case where the rent review clause provided that the tenants could serve a counternotice within three months, with time being of the essence, the tenants sought to appoint an arbitrator but it was held that in the absence of a proper counternotice and no waiver on the part of the landlord, the latter were entitled to a declaration that the reviewed rent was the figure specified in their notice. In the event it was held that there was grave hardship within the meaning of section 27 of the Arbitration Act 1950 and the tenant was granted relief (*Patel* v. *Peel Investments (South) Ltd* (1992)). A letter from the tenant of a supermarket in *Barrett Estate Services Ltd* v. *David Greig (Retail) Ltd* (1991) stated 'I note that your assessment is £190 000 per annum which I consider excessive'. The court held that as the letter was a plain statement it constituted a valid counternotice. In *Prudential Property Services Ltd* v. *Capital Land Holdings Ltd* (1993) the landlord wrote specifying a rent in accordance with the rent review clause, stating the proposed rent. By the same clause, unless tenants agreed to the proposal they were required to serve a counternotice in writing not later than one month after the landlord's notice, time to be of the essence, electing for determination by an independent surveyor. The tenant replied within the period rejecting the landlord's suggestion, proposed a lower rent and asked for negotiation, with a request to the landlord to acknowledge the letter as formal notice of disagreement. The letter was not returned but the landlord later wrote claiming estoppel since a proper counternotice had not been served. It was held that the tenant's letter was a valid counternotice since it made it clear that the tenant proposed to exercise the option.

Finally, there has been a number of cases where the court has been required to consider whether the response from the tenant constituted a valid counternotice. In *British Rail Pension Trustee Co. Ltd* v. *Cardshops Ltd* (1986) the tenant's counternotice was headed 'subject to contract'; the court determined that the proper test was whether a businessperson would be left in any doubt that the tenant was exercising his rights and that the counternotice was valid. By contrast, a letter from the landlord's agent in *Norwich Union Life Insurance Society* v. *Tony Waller Ltd* (1984) proposing a rent increase was headed 'without prejudice', it was held that the letter was equivocal and did not constitute a trigger notice. Again, in *Sheridan* v. *Blaircourt Investments Ltd* (1984) where the tenant's agents suggested an application for the appointment of an independent valuer a letter headed 'without prejudice and subject to contract' was held not to indicate sufficiently the tenant's intention to exercise his right to have the rent determined by a third party, the heading of the letter being an additional reason for the court's conclusion.

8.3 EFFECT OF LEASE TERMS

The basis for initiating and operating a rent review is provided by the terms of the lease. Some contain provisions that are simple, even naive, whilst others are of frightening complexity. Sometimes, the provisions are incomplete in which case the court may have to consider whether they are capable of interpretation or rectification. The outcome has been a sustained bout of litigation resulting in a considerable body of legal opinion, much of which is of real assistance to those concerned in managing the landlord and tenant relationship. Perhaps the important message is that the rent review provisions are worthy of careful negotiation and precise drafting in an endeavour to avoid the more obvious pitfalls.

The purpose of the rent review clause is to set out the matters to be taken into consideration and those to be disregarded in arriving at the rent payable on review. Provisions in early leases tended to follow the provisions of section 34 of the Landlord and Tenant Act 1954 which sets out the statutory basis to be followed on renewal. As landlords realised that such a basis was not always helpful to them, their advisors began to introduce bespoke provisions that not only limited the tenant's ability to negotiate freely but introduced hypothetical considerations. Typically, these sought to remove or modify any value sensitive restrictions imposed by the lease, such as narrow user clauses or onerous covenants to reinstate.

The following notes trace the main areas of difficulty and the manner of their resolution through the decided cases. Although they provide a useful guide the reader should refer to the relevant law report for full details of each case.

8.3.1 General principle

The courts will follow the general principle laid down in *Basingstoke and Deane Borough Council* v. *Host Group Ltd* (1986) that rent review clauses should be interpreted with regard to what is proper and sensible in effecting the commercial purpose unless the lease requires otherwise either expressly or by necessary implication or in a particular context.

This view was reinforced by the judgement in *Ravenseft Properties Ltd* v. *Park* (1988) where the basis of the review of land on which a supermarket had subsequently been built was to be restricted to the land without buildings, as the lease contained no assumptions and the demise was 'in consideration of the expense to the tenant of erecting a supermarket'.

A similar situation arose in *British Airways plc* v. *Heathrow Airport Ltd* (1992) where a rent review was required to find the current rental value of a building lease. It was held that the only practical approach was to assume the tenant to be under an obligation to build what had, in fact, been built. It was left for the valuer to decide whether the current tenant was likely to be a bidder for the premises and if so, at what price, taking account of its occupation of adjoining premises.

Whilst the lease provisions will govern the nature of the review it is apparent that hypothetical situations will be avoided by the courts where the lease is not sufficiently precise as to preclude the courts from arriving at a different decision. Mummery J. made clear the extent to which regard should be had to hypothetical circumstances.

> The court should also be alert to the danger of confusing reality and hypothesis. The only relevant hypotheses are those expressly or impliedly agreed by the parties in the rent review clause. Agreed assumptions and disregards must be adhered to but subject to that it is the real circumstances affecting the actual property which are relevant to valuation.

One of the questions arising regularly is whether or not the review should reflect the rent reviews provided in the actual lease. The hypothetical lease often refers to the incorporation of the terms of the lease 'other than those relating to rent'. The court has held itself entitled to select the meaning, where the language was capable of more than one meaning, which related to the apparent commercial purpose of the lease (*MFI Properties Ltd* v. *BICC Group Pension Trust Ltd* (1986)). The authorities were reviewed in *British Gas Corporation* v. *Universities Superannuation Scheme Ltd* (1986) when the literal interpretation was considered inappropriate and lead to a three part conclusion to the effect that:

1. the exclusion of all provisions as to rent would result in an outcome contrary to commercial sense that they could not be given literal effect
2. where clear words require the rent review position to be disregarded, they must be followed, even though the effect is distorting and

3. in the absence of special circumstances the words of the lease must be construed so as to give effect to the commercial purpose of the rent review clause.

Reviews were imported into the *Datastream International* v. *Oakeep Ltd* (1986) case but the courts concluded that the revised rent was to be determined on the assumption that there was no rent review clause in *National Westminster Bank plc* v. *Arthur Young McClelland Moores & Co.* (1984) and *Equity & Law Life Assurance Society plc* v. *Bodfield Ltd* (1985). A similar conclusion was reached where the revised rent was defined as 'upon the terms of this lease except as regards rent' (*Electricity Supply Nominees* v. *F.M. Insurance Co. Ltd* (1986)). In *Amax International Ltd* v. *Custodian Holdings Ltd* (1986) the landlords argued that a rent review pattern should not be inferred in a lease granted in 1970 since it was common at that time for leases to be granted without reviews (although the particular lease contained provisions for review at four-yearly intervals).

In another case the hypothetical lease was held to be on terms excluding the rent review provisions of the actual lease; the question was then the term to be assumed for the hypothetical lease which the court fixed as being the term that the landlord and tenant might reasonably be expected to grant and take at a fixed rack rent. It was further held that the assumed tenant might vary at each review. Again, the rent review provisions in a lease defined as being the rack rent without a premium with vacant possession subject to the provisions of the actual lease which was qualified by words in brackets 'other than the rent hereby reserved' was held to exclude only the rent payable (*Prudential Assurance Co. Ltd* v. *99 Bishopsgate Ltd* (1992)). Where two underleases of the same premises were granted on the same day to the same tenant, one for 16 years and the other of the expectant reversion for a further term of 34 years, the requirement of an assumption of a lease for a term equivalent in length to the term unexpired resulted in a period of one year and not the combined length of 35 years as the landlord contended (*Toyota (GB) Ltd* v. *Legal & General (Pensions Management) Ltd* (1988)).

The terms were too explicit to fall back on the general principle in *Basingstoke* (above) that the rent review clause should be construed in a commercial way. On the other hand, where the rent review clause provided that the hypothetical letting should follow the lease 'other than as to duration and rent', it was held that future rent reviews should be excluded (*Equity* above). The *British Gas* case (above) was not followed in *Philpots (Woking) Ltd* v. *Surrey Conveyancers Ltd* (1985) where the hypothetical lease was silent as to the terms or the rent review interval, the actual lease being subject to reviews at five-year intervals. Harman J. quoted Nourse L.J. in Philpots '. . . a court of construction can only hold that they intended it to have that effect if the intention appears from a fair interpretation which they had used against the factual background known to them at or before the date of the lease, including its genesis and objective aim' in refusing to ignore the terms in the actual lease as the arbitrator had done.

The hypothetical term needs to be defined carefully and with precision if the trap into which *Lynnthorpe Enterprises Ltd* v. *Sidney Smith (Chelsea) Ltd* (1990) and *Tea Trade Properties Ltd* v. *CIN Properties Ltd* (1990) fell is to be avoided. In each case there was an assumption of a letting on the same terms as the actual lease, a provision which was interpreted by the court as the unexpired residue of the existing lease. These decisions confirm the approach in *Ritz Hotel (London) Ltd* v. *Ritz Casino Ltd* (1988).

8.3.2   Basis of rent to be assessed

In the majority of cases, the intention of the review is to bring the rent in line with current values. The rent is variously described as being 'open market rental value', 'rack rental value', 'full market rent', 'reasonable market rent' and others.

It is doubtful whether there is a significant difference in these terms or that different values could be ascribed to them although in *Cuff* v. *J & F Stone Ltd* (1978) Megarry J. suggested a difference between 'reasonable rent' and 'open market rent' with these words:

> The word 'reasonable' no doubt requires the surveyor to reject a rent which though obtainable in the open market by reason of special circumstances, appears to him to exceed the rent for the premises which is right and fair: but I do not think it does more than that.

Other stipulated matters such as length of lease remaining unexpired and use will tend to affect the rental value.

Where a lease provided for a revision of the rent after five years but no formula or machinery was included in the lease for determining the rent in default of agreement, the court held that, as it would be unfair for the parties to receive no rent after the first period, there would be implied a term that in default of agreement an arbitrator should be appointed to fix the rent on the basis of a market rent (*Bates, Thomas & Son Ltd* v. *Wyndham's (Lingerie) Ltd* (1981)).

In a difference between landlord and tenant as to whether there should be an assumption of a willing tenant whether or not one existed, the Court of Appeal held that there should be no assumption that a willing tenant would pay more than is required by the market and it is a matter for the valuer to determine the rental value on the basis of the evidence available (*Dennis & Robinson Ltd* v. *Kiossio Establishment* (1986)).

Where a lease provided for the parties to agree a fair and reasonable market rent this was held to be the rent at which the demised premises might reasonably be expected to let in the open market and not a fair and reasonable rent between the parties (*ARC Ltd* v. *Schofield* (1990)).

8.3.3 Rent fixed by some other process

In some cases, there is provision in the lease for relating rent to capital value, an index or some other indicator, not necessarily resulting in the determination of a market rent. Clauses which on the face of it should be straightforward, though not necessarily equitable, because they stipulate a proportion or percentage of rack rental value as the figure to be determined for rent review purposes, have nevertheless produced uncertainty. Difficulties arose in *Stylo Barratt Properties Ltd* v. *Legal & General Assurance Society Ltd* (1989) where the tenants claimed that a percentage in excess of 100 could not be a proportion of the full rack rental value as stipulated and that no increase was due. The court found no merit in the tenant's argument.

8.3.4 Time limits

It was once thought that unless time limits contained in clauses relating to the procedure for initiating rent reviews were observed strictly, the opportunity of obtaining a higher rent would be lost. The rapid increase in rental values served only to make the landlords more anxious to receive their share. Lord Salmon understood the problem and expressed it in these terms:

> In a period of acute inflation, such as has been experienced for the last 20 years or so, and might well be experienced for many years to come, what was a fair market rent at the date when a lease was granted would probably become wholly uneconomic within a few years. Tenants who were anxious for security of tenure required a

term of reasonable duration, often 21 years or more. Landlords, on the other hand, were unwilling to grant such leases unless they contained rent revision clauses which would enable the rent to be raised at regular intervals to what was then the fair market rent of the property demised. Accordingly, it has become the practice for all long leases to contain a rent revision clause providing for a revision of the rent every so many years. Leases used to provide for such revision to be made every 10 years. Now [1977] the period is normally every seven and not infrequently every five years. It is totally unrealistic to regard such clauses as conferring a privilege upon the landlord or as imposing a burden upon the tenant.

The time limits concern the notice which has to be given by the parties to 'trigger' the review procedure, to reply to proposals made under it or to take steps to have the new amount determined where the parties are unable to agree. There are no statutory provisions relating to the content of rent review clauses and the courts are not concerned in their determination. In the event of a disagreement between the parties as to the legal interpretation of a rent review clause, the courts have a part to play: provision for reference to a third party is usually provided where the only question is one of value as is more usually the case.

(a) Where time is not 'of the essence'

Fraser L.J. examined the need to adhere to time limits laid down in the lease in *Samuel Properties (Developments) Ltd* v. *Hayek* (1972) and expressed his opinion '. . . that the equitable rule against treating time as the essence of a contract is applicable to rent review clauses unless there is some special reason for excluding its application to a particular clause'.

The case of *Bailey, C.H. Ltd* v. *Memorial Enterprises Ltd* (1973) concerned the construction of a rent review clause where it was provided that the market rental value was to be found on a certain date. The tenant contended that as it was not determined on that date (indeed it was not ascertained until some three and a half years later) it was not payable. The court held that the provision did not prevent the rent being ascertained at some later date and that the new rent was payable from the date of the rent review.

The cases of *United Scientific Holdings Ltd* v. *Burnley Borough Council* (1977) and *Cheapside Land Development Co. Ltd and others* v. *Messels Service Co.* (1977) (taken together by the House of Lords) proved to be the watershed of 'time of the essence' cases and provided

Diplock L.J. with the opportunity to review the more important of the earlier relevant decisions and to express the hope that it would reduce the number of occasions on which it would be necessary to have recourse to the courts in order to ascertain whether delay had deprived the landlord of their right.

In the *United Scientific Holdings* case, it was held that unless the lease contained any contrary indications there was a presumption that the timetable laid down in a rent review clause was not of the essence of the contract and that the review rent could therefore be determined outside the dates laid down by the lease. Salmon L.J. said:

> These [rent review] provisions as to time are not in my opinion mandatory or inflexible: they are only directory . . . A mandatory provision is one which must be fulfilled in all its strictness and failure to perform it means that the whole thing fails, whereas a directory provision does not require that degree of strictness: even though it is not complied with, the whole does not fail. It could still be regarded as valid and effective.

The *Cheapside* case contained 'an elaborate timetable' as to what was to be done in various eventualities including provisions in respect of persons over whom neither had any control whilst at some stages progress of the procedure was within the exclusive control of the landlord. The court held that the provisions for review were not of the essence and that, once fixed, the rent for the period from the review date could be recovered by the landlord.

Where a clause in a lease provided the timetable for service on the tenant of a notice to agree the rent and went on: '. . . provided always that any failure to give or receive such notice shall not render void the right of the landlord hereunder to require the agreement or determination as aforesaid of a new rent'.

Megarry L.J. was in no doubt that a notice served after the last date provided in the clause did not prevent the landlord from validly demanding agreement or arbitration of a new rent (*Kenilworth Industrial Sites Ltd* v. *E.C. Little & Co. Ltd* (1974)). One of the judges in a later case thought that the Kenilworth case was probably the origin of the unfortunate dichotomy between review clauses which conferred an option and those which merely provided machinery. In *Accuba Ltd* v. *Allied Shoe Repairs Ltd* (1975) a similar decision was given, the stipulations as to time being classified as 'mere machinery'.

In *Vince and another* v. *Alps Hotel Ltd and another* (1980) the terms of the lease for the agreement of a new rent for the second seven year period of the lease were not complied with within the times specified. Nearly two years after the review date, the landlord's attention was drawn to the *United Scientific Holdings Ltd* decision as a result of which they claimed a declaration that the review provisions could be operated retrospectively. The defendants asserted that time was of the essence and that the plaintiffs' conduct in allowing them to incur substantial expenditure in the belief that the rent would not be increased estopped them from asserting that time was not of the essence. It was held that the plaintiffs could not pursue their rights until they knew of the decision (which was a change in the law as then understood) and that there was no evidence that the defendants relied on the plaintiffs' silence. The plaintiffs obtained the declaration sought. The plaintiffs in *Dean and Chapter of Chichester Cathedral* v. *Lennards Ltd* (1977) served a notice in accordance with the terms of the rent review clause except that it did not meet the requirement that any such notice should state 'the suggested new rent reserved.' Denning L.J. remarked that the clause should not fail and be held invalid 'simply because of the omission of one bit of machinery'.

(b) Where time is 'of the essence'

One of the most helpful observations was made by Fraser L.J. when he said:

> The rule [that time not ordinarily being of the essence] would of course be excluded if the review clause expressly stated that time was to be of the essence. It would also be excluded if the context clearly indicated that that was the intention of the parties, as for instance where the tenant had a right to break the lease by notice given by a specified date which was later than the last date for serving the landlord's trigger notice. The tenant's notice to determine the contract would be one where the time limit was mandatory and the necessary implication is that the time limit for giving the landlord's notice of review must also be mandatory.

In spite of the hope expressed by Lord Diplock in the United Scientific Holdings case, activity continues on the question of the status of the time requirements of the review process.

In *Drebbond Ltd* v. *Horsham District Council* (1978) the rent review clause provided for the rent to be adjusted and if agreement was not reached a reference to a sole arbitrator '. . . by notice in writing given to the tenant within three months thereafter but not otherwise'.

The 'trigger notice', setting off the procedure, was given in accordance with the timetable but the 'arbitration notice' was not. It was held that the way in which the requirement was expressed showed the time limit to be obligatory and not merely indicative and to operate so as to make time of the essence.

A lease made lengthy and detailed provisions for rent review ending with paragraph (5): 'All stipulations as to time in the foregoing subclause . . . shall be of the essence of the contract and shall not be capable of enlargement save as agreed in writing by the parties.'

The landlord wrote in accordance with the provisions of the lease proposing a rent from the review date of £5250 per annum to which the tenant replied: 'We formally acknowledge receipt of your notice of rent review for the above property and we would hardly need to add that we do not accept your revised figure.'

No further steps were taken by the tenant until after the time limit set for agreement or service of a counternotice had expired. It was held that the tenant's letter contained nothing sufficiently specific to constitute a counternotice and judgement was given in favour of the landlord (*Bellinger* v. *South London Stationers Ltd* (1979)).

A similar case was *Oldschool and another* v. *Johns* (1980) where the lease provided that in the absence of agreement, a counternotice was to be served by a certain date requiring determination of value by an independent surveyor (time to be of the essence thereof). A letter from the tenant's solicitors disputing the rental value and asking for evidence in support of the landlord's proposed rent was held to be at the best ambiguous and in any event not an effective counternotice. It was further held that certain actions of the landlord did not constitute waiver or estoppel.

In *Amalgamated Estates Ltd v. Joystretch Manufacturing Ltd* (1980) it was held that a landlord's notice specifying a new rent need not be a bona fide and genuine pre-estimate of the rent and that a tenant's reply to the notice in which he disagreed with the proposed increase and made a request for an explanation of the basis of the landlord's proposal was not a counternotice. All the stipulations were subject to time being of the essence.

It was pointed out that in such cases it is possible for the tenant to apply to the High Court for an extension of time for the commencement of arbitration proceedings which may be granted where the court is of opinion that undue hardship would otherwise be caused.

Another case where time was stated to be of the essence was *Weller* v. *Akehurst* (1980) where it was held that when the provisions for ascertaining open market value were defined with precision and were required to be effected in a particular way, the provision as to time being of the essence defeated the landlord. Having failed to observe the timetable, he was unable to obtain a review to the open market value. In these circumstances it was held that the original rent should continue. But in *Bates, Thomas & Son Ltd* v. *Wyndhams (Lingerie) Ltd* (1981) where the lease provided no machinery for fixing the rent on review it was said that there was no difficulty in implying a term that a market rent should be fixed.

Other cases of interest include *Bradley, C. and Sons Ltd* v. *Telefusion Ltd* (1981) where the lease provided that time was of the essence but it was uncertain as to whether it applied to the whole of the clause or only to parts of it. It was held that it was intended that time should be of the essence for the arbitrator to make his determination and, as he did not do so, the landlord was out of time.

In *Edlingham Ltd* v. *MFI Furniture Centres Ltd* (1981) time was again stated to be of the essence in a rent review clause. In reply to the landlord's notice specifying a new rent in accordance with the clause the defendant company's company secretary acknowledged receipt of the notice, stated that the suggested rent was considered excessive and requested the comparables on which it was based. It was held that the letter was not effective as a counternotice and therefore the new rent was properly due (and see *Oldschool* above). Again, there was some discussion of the possibility of the tenant applying under section 27 of the Arbitration Act 1950 for leave to proceed despite the lapse of time: however, the aggrieved party is required to persuade the court that in the circumstances of the case undue hardship would be caused if leave was withheld.

A further question of construction was raised in *Al Saloom* v. *Shirley James Travel Service Ltd* (1981). The case was concerned with whether the landlord had served notice in accordance with the requirements of the clause in the lease which contained provisions about break and review procedures. It was held that as the two provisions were closely allied to one another, the same interpretation must apply to each. Diplock L.J. pointed out that there was a practical business reason for treating time as of the essence of a break clause. It was held that the phrase must mean the same in relation to each clause and therefore that time must also be treated as of the essence of the review clause.

In *Rahman and others* v. *Kenshire* (1981) the rent review clause and a break clause were interrelated: there was no question but that time was of the essence in the break clause and again it was held that it must also be of the essence in the review clause. The intention of the document appeared to be to give the tenant an opportunity to determine the tenancy after the review rent was known: the intention would not be achieved unless both clauses were treated in the same way. Break clauses and options to determine on the part of the landlord must be pursued through a notice under the Landlord and Tenant Act 1954 which is considered in Chapter 7.

Despite the general rule that time is not of the essence, the professional advisor is well counselled to comply with any timetable laid down in the rent review clause. The courts have expressed disquiet from time to time about the operation of a strict timetable but the particular wording of a clause may lead to the inescapable conclusion that time is of the essence.

In *Factory Holdings Group Ltd* v. *Leboff International Ltd* (1986) it was found that even though the rent review provisions did not make time of the essence it was open to one of the parties to do so if in the circumstances it was a fair and reasonable action. It was held that it was not in this case where the tenant required the landlord to refer the matter to arbitration within 28 days as the tenant could himself have initiated the arbitration process. Similarly, where a landlord was dilatory in referring a matter to arbitration where he had the ability to do so but time was expressly stated not to be of the essence, it was observed in *Panavia Air Cargo Ltd* v. *Southend-on-Sea Borough Council* (1988) that the tenant was able to serve a notice for determination and for this purpose make time of the essence.

In *Power Securities (Manchester) Ltd* v. *Prudential Assurance Co. Ltd* (1986) Millett J. acknowledged the difficulty in reconciling decisions on time of the essence and, in response to a claim by the landlord that time should be of the essence in the provision by the tenant of a schedule of income received from the demised premises, set out his opinion of the principles to be derived from decided cases:

1. The correct approach is to begin with a presumption that time is not of the essence in the rent review procedure.
2. Accept displacement of the presumption where the conditions of the lease as a whole and in particular the provisions of the review clause as a whole show a contrary intention.
3. Where the parties have required a step to be taken within a specified time and provided for the consequences on default, this provides an indication of intention but is not to be regarded as decisive which must depend on all the circumstances of the case.
4. In the end the matter is one of impression, giving effect to the lease and the rent review clause but avoiding fine distinctions.

Whilst accepting that it was necessary that a time limit should be fixed it was insufficient to imply that time was of the essence, since the provision of what was to happen in the case of default was a sufficient indication in this case that time was not to be of the essence.

In *Leeds Permanent Pension Scheme Trustees Ltd* v. *William Timpson Ltd* (1987) where there was a detailed timetable by which certain events should have taken place the court nevertheless held that there was nothing in them to overturn the general principle that time was not of the essence. By contrast, in the case of *Thorn EMI Pension Trust Ltd* v. *Quinton Hazell plc* (1983) the court held that the expression '... in any event not later than 4 months ...' was a sufficient contraindication to displace the general rule.

The effect of a tenant's option to determine the lease continues to support the decision in *Coventry City Council* v. *J. Hepworth & Son Ltd* (1981) and *William Hill (Southern) Ltd* v. *Govier & Govier* (1983) that time is of the essence since it is important for the tenant to know the level of the rent to be fixed before committing himself to a further term by not exercising the option. Where a notice was to be served by a certain date which happened to be a Sunday and the recorded letter was not delivered until the following day by which time it was out of time ensured that the tenant had insufficient time to comply with a further provision of the lease if he wished to exercise his right to break; the decision in this case, *Sterling Land Office Developments Ltd* v. *Lloyds Bank plc* (1984), was that time was of the essence following *Al Saloom* v. *Shirley James Travel Service Ltd* (1981). *Vision Hire Ltd* v. *Britel Fund Trustees Ltd* (1992) confirmed that the decision in *United Scientific Holdings* was applicable to Scotland. A condition precedent of the landlord's right to a rent review was a notice in writing from him within a specified timescale lead to a decision in *Chelsea Building Society* v. *R & A Millett* (Shops) *Ltd* (1993) that time was of the essence. *Art & Sound Ltd* v. *West End Litho Ltd* (1992) is a further example of a situation where part of the process, in this case the arbitrator's decision, may be subject to time of the essence where the lease makes the situation clear.

There is some evidence that the elaborate clauses designed to avoid the various pitfalls of the rent review arena have not been entirely successful.

8.3.5 Fitness for use

Rent-free, concessionary and fitting-out periods are more prevalent where a difficult letting market results in initial incentives to secure occupation by a tenant. Rent review clauses intended to exclude the effect of rent-free concessions or fitting-out periods were considered in *City Offices plc* v. *Bryanston Insurance Co. Ltd* (1993). The court supported a contention that if no rent-free period was available, a lower rent would be agreed to reflect that. A similar result was achieved in a case concerning another tenant of the same landlord (*City Offices plc* v. *Allianz Cornhill International Insurance Co. Ltd* (1993)).

The question of fitness for immediate use was raised in *London & Leeds Estates Ltd* v. *Paribas Ltd* (1993) where the disregards were directed towards preventing the tenant from claiming a discount for time to carry out the work.

Pontsarn Investments Ltd v. *Kansallis–Osake–Pankki* (1992) helpfully defines the meaning of several phrases, at least in relation to that case but also of wider application. Thus 'vacant but fit for immediate use and occupation' meant that the building was ready for the tenant to take possession to fit it out to enable him to commence trading and 'ready for occupation' meant that the building was available for fitting-out purposes.

It seems evident that many of the drafting attempts to avoid the effect on rent reviews of the various discounts now on offer will prove lacking.

8.3.6 Restrictions as to user

Where the lease limits the user in some way, the rent must be assessed with regard to that restriction which will often (but not always) produce a lower rent for the landlord.

In one case there was a restriction on use to offices in connection with the lessee's business of consulting engineers. The arbitrator was urged to take into account a possible relaxation of the terms of the user clause. He stated a case for the court which held that he must make an award on the basis of the user clause and not any variation which either party might be prepared to agree to (*Plinth Property Investments* v. *Mott, Hay and Anderson* (1979) where the rent was found to be £130 355 per annum without the user restriction and £89 200 per annum with the restriction).

In the course of his judgement Brandon L.J. said in relation to the contention that the landlord would be prepared to relax the restrictions:

> What the arbitrator has to consider is what those rights and obligations are on either side and assess the rent in the light of them.

He is not to say to himself 'Those who have rights may not enforce them and those who have obligations may not be required to enforce them'. He is to assume that the rights will be enforced and the obligations will be performed. He is to look at the legal position of the parties and nothing else.

The Court of Appeal had no hesitation in ruling that a possible future use in breach of planning control was not a matter to be taken into account when assessing the review rent. But where the review rent was to be assessed on the basis of the open market rental ignoring any statutory restriction, it was held that the flats included in the lease were to be valued disregarding the possibility of the rents being subject to a ceiling fixed by the Rent Officer (*Langham House Developments Ltd* v. *Brompton Securities Ltd* (1980)).

An unusual lease provision was considered in *Trust House Forte Albany Hotels Ltd* v. *Daejan Investments Ltd* (1980) where it was provided that parts of the demised premises used as a hotel were to be valued as if they were let for retail purposes. The landlords contended that the clause required the further assumption that the premises were in a state suitable for such use but this contention was not upheld.

In *Wolff* v. *Enfield London Borough* (1987) the rent review clause was widely drawn as to the use to be assumed '... for any purpose within Class III of the Town and Country Planning (Use Classes) Order 1972 or any other class of the order permitted by the local planning authority from time to time'. Nevertheless, use by an educational institution for a non teaching purpose could not be taken into account since the use authorised by the planning permission was a composite use not falling within any of the prescribed use classes. The rent was to be assessed therefore on the basis of use for light industrial purposes within Class III.

Any restriction on use is potentially damaging to the rental value of the premises and is a constant source of litigation. In *Sydenham, J.T. & Co. Ltd* v. *Enichem Elastomers Ltd* (1986) the tenant claimed that the user clause which restricted the use to the business he carried on would affect the rental value; it was held that this was not so, since the lease authorised other uses and the particular clause would not therefore bind an assignee. A letter from a landlord purporting to authorise the tenants to use the demised premises for a range of uses which was not given at their request and not needed by them did not enable the landlord to use the permission to justify a higher rent on review *C&A Pension Trustees Ltd* v. *British Vita Insurance Ltd* (1984). This only confirms the conclusion reached in earlier cases, notably *Clements, Charles (London) Ltd* v. *Rank City Wall Ltd* (1978).

A user clause personal to the tenant was held not to preclude the use of the premises by others because although consent would be required the lease provided that it could not be unreasonably withheld (*Mars Security Ltd* v. *O'Brien* (1991)). A user clause for the purposes of the hypothetical letting that limited the use of the premises to that of a branch post office and an absolute restriction on assignment justified the arbitrator's approach in *Post Office Counters Ltd* v. *Harlow District Council* (1991).

A user provision to the effect that the premises should be occupied only as a branch of Lloyds Bank in *Sterling* (above) was held to be a restriction to the user of the hypothetical willing lessee, the name of that hypothetical tenant to be inserted when known, following the precedent set by *Law Land Co. Ltd* v. *Consumers' Association Ltd* (1980). A provision in the rent review clause required the rack rental value to be determined on the basis that the premises were let for office purposes in *Bovis Group Pension Fund Ltd* v. *G.C. Flooring & Furnishing Ltd* (1984). The Court of Appeal held that it should be assumed that the premises would be used lawfully for that purpose.

On the question of a limited user covenant, where the assignee had been granted use for another purpose, it was held that the user specified in the lease was the basis of the use for rent review purposes and was unaffected by subsequent arrangements made in a licence for the benefit of the current tenant (*SI Pension Trustees Ltd* v. *Ministerio de Marina de la Republica Peruana* (1988)). A similar conclusion was reached in *Postel Properties Ltd* v. *Greenwell* (1992) where the actual use permitted by the lease differed from the specific provision for a wider user under the hypothetical lease.

8.3.7 Provisions as to unexpired term

In an endeavour to avoid the possibility of lower rental values being fixed as a result of the limited term available, some leases seek to overcome the problem by substituting specific assumptions as to the expired term, overriding the actual position.

In *Pivot Properties Ltd* v. *Secretary of State for the Environment* (1980) the lease provided for the 'rack rental market value' to be fixed on review, being 'the best rent at which the demised premises might reasonably be expected to let in the open market for a term not exceeding five and a half years'.

The arbitrator made alternative awards: if any extension available under the Landlord and Tenant Act 1954 was to be reflected, the rental value was £2 925 000, otherwise it was £2 100 000. The Court held that the possibility of extension was to be taken into account.

8.3.8 Unexpired term

In the circumstances of an unusual lease provision where the rent was to be reviewed over one year after the commencement of the lease and the rent then to become the rent payable from the beginning of the lease, it was held in *Prudential Assurance Co. Ltd* v. *Gray* (1987) that the 'date of review' was the date on which the new rent was actually determined.

8.3.9 Assumption of vacant possession

It is common to provide that vacant possession is to be assumed when assessing rental value on review. In the recent case of *Avon County Council* v. *Alliance Property Company Ltd* (1981) the Court looked at the background to the original lease (even though both parties had been superseded) and concluded that the rent was related to land and buildings and not to the leasehold interests and that it was possible to imply a vacant possession basis even though such a stipulation was not contained in the head lease and the property was subject to a number of subleases and licences.

8.3.10 Treatment of tenant's improvements

The status of the tenant's improvements on renewal is made clear by the 1954 Act: subject to certain qualifications such improvements are ignored by the court in fixing the rent under the new lease. On review, however, the intention of the parties will suggest the basis and that intention will be found primarily in the lease. Two interesting cases on this point are *Cuff* v. *J. and F. Stone Ltd* (1978) and *Ponsford* v. *HMS Aerosols Ltd* (1978).

In *Cuff* the tenants carried out some alterations to increase the floor area during the currency of their 21-year lease. The tenants obtained a statutory renewal at the end of that lease for the maximum term of 14 years and a rent review was incorporated to operate at the end of the seventh year. The rent under the new lease ignored the rental value of the tenant's improvements but when the rent fell to be reviewed the landlord claimed an amount to include the value of the improvements. He was held to be entitled to a rent on that basis as the review clause referred to 'a reasonable rent for the demised premises' the point being that the improvements carried out by the tenants had become part of those premises and were therefore, in the absence of a specific provision in the lease, to be included in the assessment of a reasonable rent.

In the *Ponsford* case, a serious fire occurred a year after a 21-year lease had been granted to the tenant. The premises were rebuilt and the landlord agreed to substantial improvements being incorporated by the tenant at his expense – the not inconsiderable sum of £31 780. Within five years of rebuilding a review took place and the landlord claimed a rent to include the rental value of the improvements. The court upheld the landlord's claim and drew attention to *Cuff* (previously unreported). It is clear that the statutory disregard in section 34 cannot be imported into a review clause, the presumption being that the parties strike the bargain they intend. The case was decided by a 3–2 majority in the House of Lords. While it is difficult to see how any other decision could have been reached on the facts, it is doubtful whether such a situation was intended by the legislators. It is ironic to reflect that when the lease ended in 1989, the tenant was in a position to claim a new lease at a rent which disregarded the effect of his improvements.

A difficult case of tenant's improvements was considered in *GREA Real Property Investment Ltd* v. *Williams* (1979). The tenant took a bare shell being the third floor of an office block and proceeded to carry out all the remaining building works including partitions, lavatories, cloakrooms and so on. The first six months of the lease was at a peppercorn. The review clause provided for the disregard of any effect on the rental value of any improvement carried out by the tenants, their works of fitting out and finishing being referred to specifically in this exclusion. On review, the landlord's valuer estimated the rental value of the completed suite and deducted a percentage for the tenant's improvements while the tenant's valuer deducted the annual equivalent of the updated cost of the improvements from the rental value. As a result of a consultative case from the arbitrator, the judge advised him – among other things – that the intention of the parties had been to credit the tenant with the rental equivalent of the updated cost but with nothing else. In the course of his consideration he made some helpful general points of interest to valuers, including a reference to the difference between cost and value though he refrained from expressing a view as to the valuation method to be used in arriving at the rental value.

The same judge had this to say in a later case:

> Now this court is not a valuer. All I can do is to say whether there is an error on the face of the award and there will be such an error if a method of valuation is adopted which clearly does not follow the intention of the parties. I can say, therefore, whether a method of valuation is wrong, but not necessarily what method of valuation is right.
>
> Forbes J. in *Estates Projects Ltd* v. *Greenwich London Borough Council* (1979)

Following Ponsford where the tenant was held by a majority decision of the House of Lords to be liable to pay rent on improvements paid for by him, there was a similar outcome in *Laura Investment Co. Ltd* v. *Havering London Borough Council* (1992). A long lease of undeveloped land was granted subject to a covenant to divide the land and underlet the plots for the construction of buildings. Failing any specific provision in the lease (as in *Tesco Holdings Ltd* v. *Jackson* (1990) and *Hill Samuel Life Assurance Ltd* v. *Preston Borough Council* (1990)), the reviewed rent fell to be assessed to include the buildings. By contrast, a rent review of a football ground was held to be restricted to the ground itself, excluding the buildings thereon, because the covenants in the lease made a distinction between the ground and the buildings (*Ipswich Town Football Club Co. Ltd* v. *IBC* (1987)).

An important case, *Lear* v. *Blizzard* (1983), had facts which enabled the judge to distinguish *Ponsford* v. *HMS Aerosols Ltd* (1978). He held that the rent should be a fair one, between this particular landlord and this particular tenant and that the improvements should be discounted wholly or partly according to the tenant's contribution to the work.

8.3.11 Restrictions on assignment

Leases often contain an absolute or conditional prohibition of assignment of the unexpired term of the lease.

The precise effect of the restriction on rental value needs to be considered in the light of the practical effect on the marketability of the lease. A clause requiring tenants to offer to surrender the lease to their landlord before proceeding with any assignment to a third party was held to be contrary to section 38 of the Landlord and Tenant Act 1954. This was so, even though there were provisions for the landlord to match the payment offered by the prospective assignee (*Allnatt (London) Properties Ltd* v. *Newton* (1980)).

8.3.12 Profitability of tenant's business

Tenants cannot be required to produce their books of account under normal circumstances, a proposition confirmed by *Barton, W.J. Ltd* v. *Long Acre Securities Ltd* (1982). However, it was held in *Harewood Hotels Ltd* v. *Harris* (1957) that evidence of accounts was admissible in showing the reasonable open market rent likely to be achieved whilst in *St Martin's Theatre, in re: Bright Enterprises Ltd* v. *Lord Willoughby de Broke* (1959) it was held that the tenant could be required to produce his accounts.

8.3.13 Delay

Where there is a delay in determining the rent under a rent review it is clear that the landlord can recover interest on the balance only where the lease itself makes specific provision for payment in such circumstances.

In *Parry* v. *Robinson-Wyllie Ltd* (1987) the determination of the rent review was much delayed during which time a receiver was appointed by the tenant company, the defendant company having taken an assignment of the lease only a month after the new rent was determined. The landlords claimed the arrears from the defendant company which was held to be not liable, although it was found that the receiver had a liability. Where an application to the President of the Royal Institution of Chartered Surveyors to appoint an independent surveyor to determine the rent in a period after that provided for in the rent review clause it was held in *Darlington Borough Council* v. *Waring & Gillows (Holdings) Ltd* (1988) that the procedure had not been properly invoked and the rent for the ensuing five years would not therefore be changed.

8.3.14 Miscellaneous

There are other matters of interest to the property manager. It has already been noted that decisions support the conclusion that interest on rent paid late following arbitration cannot be claimed unless specifically provided for in the lease.

The uncertainty surrounding the licensing of stalls in ancient markets and in particular the level of rent and the way in which it was calculated all received consideration in *R. v. Birmingham City Council ex parte Dredger* (1993) where the council's decision to more than double some rents was the subject of judicial review. The outcome is a landmark for traders of municipal space lacking the protection of a lease and the judgement makes it clear that the authority must act reasonably. Where the tenant granted exclusive possession to individual market traders in an enclosed hall, it was held that it could not claim protection of the act against the landlord since the tenant did not occupy the stalls for the purpose of its business and neither could it be said that it occupied the common parts alone.

The issue of doctrine estoppel exists to prevent relitigation (as opposed to appeal) of a particular point. In *National Westminster Bank plc v. Arthur Young McClelland Moores & Co.* (1984) the court determined that the hypothetical lease should be construed as not containing any rent review (which had the effect of increasing the rent to be assessed by the arbitrator). When a later decision in another case confirmed the mistake in this judgement the lessees attempted to have the lease rectified by the court or a declaration as to the true construction of the rent review clause. A certificate was therefore granted under section 1(7)(b) of the Arbitration Act but there was no change from this course. The tenants made a further attempt in 1990, *National Westminster Bank plc v. Arthur Young McClelland Moores & Co.* (1990), but it was held that only one application could be made under the section and it had to be within the appeal limits.

8.4 THIRD PARTY DETERMINATIONS

Any provision for the review of rent during the course of a lease should anticipate that there will be times when the parties cannot reach agreement on the new rent. There is a need for the lease to contain machinery to enable the dispute to be resolved and it is normal to provide for the appointment of a third party, usually a surveyor, to act as arbitrator or expert (sometimes referred to as independent valuer).

Where an arbitrator is appointed they will conduct the proceedings in a formal manner much in line with court procedures. Where a point of law is at issue they may, with the agreement of the parties, sit with a legal assessor or seek legal advice: one of the parties may seek a direction from the court.

The statutory background to arbitration provides an opportunity for either party to have recourse to the High Court on a point of law or where it appears that the operation of the procedure or the result is irregular. Some of the more important issues and the associated results of court actions will be mentioned later. They may take evidence orally or in writing: evidence is likely to include valuation calculations and analyses of comparables. The evidence may be tested by cross-examination and re-examination and it is usual for the arbitrator to inspect the subject property and the comparables as far as is possible.

Where required by the parties or by the instrument appointing them, their award must be reasoned. They may find it appropriate to make alternative awards where the result turns on a point of law. Unless excluded by agreement the parties have a right of appeal to the High Court on a matter of law subject to obtaining leave of the court.

Fees may be fixed by agreement with the arbitrator when they are first appointed or may be based on the amount of their award. Fees may fall equally on the parties, follow the award (so that the unsuccessful party is responsible for payment) or be shared in the discretion of the arbitrator. Either party may apply to have the arbitrator's fees taxed by the court.

In recent times there has been a tendency for the expert to replace the arbitrator. Unlike arbitrators, experts are expected to use their expert knowledge and are not bound to hear the parties unless they reserve the right to appear.

Arbitrators must conduct themselves in a judicial manner and may be removed or have their award remitted or set aside by the court where any irregularity is shown. Some examples follow.

8.4.1 Errors of law

In *Triumph Securities Ltd* v. *Reid Furniture Co. Ltd* (1986) the arbitrator's award was challenged on the ground that he stated as a rule of law that a specific percentage should always be deducted to allow for the relationship of frontage to depth in ground floor premises but the court held that the deduction made was his view of what was appropriate in this case and was not intended to suggest a general rule.

The arbitrator in *Broadgate Square plc* v. *Lehman Brothers Ltd* (1993) reflected the effect of an inducement (although he had received correct advice from his legal assessor based on the terms of the rent review clause) disclosing an error in law.

In *Arnold* v. *National Westminster Bank plc* (1993) an application to remit an award was based on an error of law disclosed by later decisions of the courts but was refused because of the delay in making the application. The proper test to apply in granting leave to appeal on a question of law was considered in *Ipswich Town Football Club Co. Ltd* v. *IBC* (1987). It concluded that the courts should approach the question with a bias towards finality and that a strong prima facie case needed to be made out which it was in this case.

An appeal on a question of law succeeded in *Prudential Assurance Co. Ltd* v. *99 Bishopsgate Ltd* (1992) where the rent review clause required that the yearly rent was to be fixed on the basis that the property was available to let with vacant possession. The landlord contended that the arbitrator was to assume the building was already let. The court varied the arbitrator's award to reflect the rental value with vacant possession.

The arbitrator was challenged on issues of law and fact in *Amego Litho Ltd* v. *Scanway Ltd* (1994) and was removed on the ground that he had rejected a piece of evidence from the tenant.

The question as to whether an arbitrator was entitled to depart from an assumption agreed between the parties was considered in *Techno Ltd* v. *Allied Dunbar Assurance plc* (1993). As the arbitrator had not approached the parties for their views he was guilty of misconduct and the case was remitted.

In a case where the parties' surveyors put forward valuations on two different bases that for the tenant on a profits-type approach, the arbitrator criticised the valuations as unsubstantiated and unreliable. The tenant's challenge was unsuccessful, the court holding that there was no obvious error of law in the arbitrator's approach (*Temple & Crook Ltd* v. *Capital & Counties Property Co. Ltd* (1990)).

8.4.2 Findings of fact

In *My Kinda Town Ltd* v. *Castlebrook Properties Ltd* (1986) the tenants alleged errors of law in the award but the court held that the arbitrator based his decision on two findings of fact from which there was no appeal. An application for leave to appeal from an award was made in *Segama NV* v. *Penny le Roy* (1983) on the grounds that the arbitrator had admitted evidence of rents agreed after the date of the rent review. The court regarded the matter as one for the arbitrator.

8.4.3 Admissibility

It was held in *Land Securities plc* v. *Westminster City Council* (1992) that an arbitrator's award is his opinion based on evidence given before him which is hypothetical rather than real and is not therefore admissible in support of rents proposed in another rent review dispute.

8.4.4 Extension of time

Section 27 of the Arbitration Act 1950 confers on the court a right to grant an extension of time which was confirmed in *Pittalis* v. *Sherefettin* (1986) even though the county court judge had initially given judgement for the lessors and then recalled that judgement and found in favour of the tenants. The court granted an extension of time for the service of a rent review counternotice by the tenant even though the substantial delay was unexplained *Patel* v. *Peel Investments (South) Ltd* (1992).

In the case of *Richurst Ltd* v. *Pimenta* (1993) the landlord started the rent review procedure by serving a notice three weeks out of time and applied for an extension under section 27 to regularise the matter. It was held that the court had no jurisdiction since the landlord's notice was not part of the arbitration provisions and the service of the notice not therefore 'a step to commence arbitration proceedings'.

8.4.5 Remission

After offering an oral hearing at the behest of either party which was taken up by the tenant's surveyor the arbitrator proceeded to issue his award without a hearing and the award was remitted to another arbitrator (*Sotheran (Henry) Ltd* v. *Norwich Union Life Insurance Society* (1992)).

In *Unit Four Cinemas Ltd* v. *Tosara Investment Ltd* (1993) the arbitrator's award stated that he had made his determination on a profits basis whereas the parties had submitted evidence on turnover and comparison bases. The award was remitted for reconsideration by the arbitrator.

An application for an award to be remitted in *Secretary of State for the Environment* v. *Reed International plc* (1993) was not granted where the arbitrator showed a small uplift in rental value per square foot on the basis that there would have been an active bidder in addition to the hypothetical tenant. The reasons given by the arbitrator were described by the court as the exercise of pure valuation experience.

Section 22 of the Arbitration Act 1950 allows remission of an arbitrator's award in certain circumstances. In *Arnold* (above) the tenant sought an order on the grounds that the arbitrator's award contained an error of law as evidenced by later decisions of the courts. This particular application was refused on the ground that there was excessive delay in making the application.

8.4.6 Misconduct

Arbitrators can be removed by the court if they misconduct themselves either during the course of the arbitration procedure or in their award.

In *Control Securities plc* v. *Spencer* (1988) the arbitrator failed to send a copy of the tenant's letter to the other side and also failed to hold an oral hearing which he had undertaken to discuss the need for in his letter setting out rules of procedure. The challenge was successful.

An attempt by the tenant to have the court order the arbitrator to stand down where he had been involved in an award which the tenant wished to quote was unsuccessful (*Moore Stephens & Co.* v. *Local Authorities Mutual Investment Trust* (1992)).

In *Handley* v. *Nationwide Anglia Building Society* (1992) the judge expressed his opinion that the arbitrator should have given the parties an opportunity to comment on his intention to make certain reductions from comparable evidence submitted and that he should not have come to a conclusion about a location on the basis of his own expert opinion: the award was set aside.

The advantages claimed for the appointment of an expert are that the procedure is quicker and less cumbersome and therefore cheaper: that it is preferable, where the parties' representatives have been unable to resolve the issues, to have a completely fresh approach and that the expert is liable to the parties in negligence. Table 8.1 sets out the advantages and disadvantages of each approach.

When the lease is entered into, the precise basis of any future dispute as to the rent payable on review is a matter of conjecture and the decision as to whether to provide for the appointment of an arbitrator or independent valuer is without any firm basis. Such advantages as there are in the appointment of an expert should be weighed against the detailed legal framework for arbitration set out in the Arbitration Acts 1950 and 1979.

8.4.7 Developments in the approach to settlement of disputes

There has been some concern at the delay in receiving the arbitrator's award and the level of costs involved. There are two developments

Table 8.1 Comparison of the advantages and disadvantages of arbitrators and experts

	Independent valuer	Arbitrator
Requirement	to use their knowledge and expertise in resolving the dispute	to take evidence and to make an award having regard to the evidence and within the parameters presented by the parties
Right of parties to make written representations	not unless agreed by parties and made a condition of appointment	determined by arbitrator in accordance with provisions of the Arbitration Acts 1950 and 1979
Right to a hearing	not unless agreed by parties and made a condition of appointment	determined by arbitrator in accordance with provisions of the Arbitration Acts 1950 and 1979
Prescribed procedure	not unless agreed by parties and made a condition of appointment	determined by arbitrator in accordance with provisions of the Arbitration Acts 1950 and 1979
Discovery of documents	no power to order	determined by arbitrator in accordance with provisions of the Arbitration Acts 1950 and 1979
Attendance of witnesses	no power to order	determined by arbitrator in accordance with provisions of the Arbitration Acts 1950 and 1979
Power to re-appoint on incapacity	no	determined by arbitrator in accordance with provisions of the Arbitration Acts 1950 and 1979
Appeal on point of law	only where reasons or calculations are given and can be shown to be wrong	
Award of costs	no	enforceable as a judgement debt
negligence	liable	not liable
Speed	depends on nature and complexity of dispute: independent determination is often promoted as being quicker.	
Cost	depends on nature and complexity of dispute: independent determination is often promoted as being cheaper.	

that have yet to gain a firm hold but that are worthy of consideration in attempting to reach a quicker and cheaper resolution of the dispute.

The first is the so-called pendulum or flip-flop arbitration where the arbitrator considers the evidence and then chooses the valuation submitted by one of the parties; it is one or the other – there can be no intermediate award.

It is suggested that this form of resolution concentrates the mind; if either party goes beyond what is reasonable there is a real chance that the other party's reasonable proposal will be the one selected. Where both parties are acting reasonably there will be a greater convergence of result with the loser being within a more acceptable range of outcomes.

The second is known as alternative dispute resolution (ADR). The parties appoint a neutral mediator whose role is to give guidance in helping the parties to negotiate a settlement. Mediators have no powers and make no award; their role is performed by helping the parties to explore their positions and define the principal points to be resolved. Confidential information received from one party will be disclosed to the other only if permission is given. The process could be undertaken within a tight timescale and, because of the limited role of the mediator, the costs should be considerably lower than the traditional methods of resolving disputes. It is unlikely that either alternative will be readily or quickly adopted by the parties or their professional advisers but they are worthy of consideration in appropriate cases.

8.5   IMPLEMENTATION

More often than not the property manager or solicitor serves the appropriate notice to commence the review procedure, the former then proceeding to negotiate the rent payable on review. Although there has been a flood of litigation on this subject the vast majority of reviews has been concluded, if not amicably, at least without recourse to the courts or to third party proceedings.

When agreement has been reached on the review rent it amends one of the terms of the lease and the parties' solicitors should take the necessary action: as there is no intention to create a new tenancy the terms of the agreement may be endorsed on the existing lease. A new lease is unnecessary. The point is important: an endorsement noting an agreed variation in the terms of an existing lease is not a lease for stamp duty purposes but a bond or covenant subject to section 64(i)(a) of the Finance Act 1971.

FURTHER READING

Brahams, D. and Pawlowski, M. (1986) *Casebook on Rent Review and Lease Renewal*, Collins.
Clarke, D.N. and Adams, J.E. (1990) *Rent Reviews and Variable Rents*, Longmans Law, Tax and Finance.

Residential tenancies

9

Lettings in the residential sector have been typified by strict statutory security of tenure coupled with artificially-controlled rents. The Housing Act 1988 broadly ended rent control and reduced the difficulties in regaining possession. Subsequently the Business Expansion Scheme gave a boost to the private rented sector in providing a vehicle for investment, tax saving and subsequent realisation to show a profit. Although the subsequent recession has upset these plans, the two events have resulted in a greater supply of better type dwellings becoming available to let. As a result, the management of dwelling houses has been revived, bringing with it a need for an understanding of current law in this area if the major pitfalls are to be avoided. Tenants now enjoy greater safeguards in the assessment of service charges; certain tenants of long leasehold premises at low rents have rights to buy or to gain extensions to existing leases.

It is intended to describe the two codes at present in force, highlighting the extent to which the parties remain subject to external (statutory) direction. The responsibility for repairs, the assessment of service charges and provisions for subletting, premiums and notices will follow and the chapter will end with short notes on the enfranchisement of long leasehold interests.

9.1 INTRODUCTION

The Housing Act 1988 is the latest in a long line of statutes, enacted over the past 65 years. Initial steps to control rents and provide security of tenure were taken in 1915, early in the First World War, when most families rented their houses from private landlords. The legislation has been prolific, complicated, often partisan and largely unsuccessful.

The purpose of the Rent Acts was well described by Scarman L.J. in *Horford Investments Ltd* v. *Lambert* (1974) where he said:

The policy of the Rent Acts was and is to protect the tenant in his home, whether the threat be to extort a premium for the grant or renewal of his tenancy, to increase his rent or to evict him. It is not a policy for the protection of an entrepreneur such as the defendant whose interest is exclusively commercial, that is to say, to obtain from his tenants a greater rental income than the rent which he has contracted to pay his landlord. Throughout their history the Rent

Acts have constituted an interference with contract and property rights with the specific purpose of redressing the balance between landlords and those who rent their homes. Social engineering of this kind is properly a function of the state and the political interference has understandably made investors wary of providing housing to rent.

9.2 RECENT DEVELOPMENTS

Today, this is less true. Any tenant of residential premises is the subject of one of two codes.

Most residential tenancies are subject to some form of direction, the principal current relevant legislation being contained in the consolidating Rent Act 1977 and the Protection from Eviction Act of the same year and the Housing Act 1988.

Tenancies within the scope of the earlier code give an occupier protection either as a contractual or more likely a statutory tenant who, as a regulated tenant is entitled to control of the rent payable and security of tenure. The landlord can reclaim possession only on proof of one of a number of available mandatory or discretionary grounds to the satisfaction of the county court.

The thrust of more recent legislation, particularly the Housing Act 1988, has been to reduce the degree of control, particularly on the landlord's right to obtain possession when required. Lettings subject to the Housing Act 1988 have only limited security of tenure, assured shorthold tenants having none. Also it has removed the wider right of a family member to succeed to a statutory tenancy, limiting the claim to succeed to a statutory tenancy to the surviving spouse. The latest legislation may signal the beginning of a more successful approach than hitherto, where successive governments have failed to achieve even modest success in the housing field.

In the newly-emerging, post-boom society where numerous families have lost the homes which were intended to give them security and a higher standard of living, some have negative equity and others wish to sell for economic or employment reasons and find the market unaccommodating. Whilst the situation is hurtful to many, it may result in the acquisition of accommodation to be approached on a more rational basis for the primary reason of shelter, where the conclusion might be that a rented house would be more appropriate to their particular circumstances. This may point to an opening for the provision of houses to let. Housing

Associations have offered the opportunity for some time but are increasingly caught between the need to have regard to their social obligations and the increased cost of capital, now that funds have to be raised partly on the commercial market.

The residential housing market is the only investment area where the investor has been forced through restrictive legislation to accept an uncertain and inadequate return on capital coupled with, for the most part, imposed and extended contracts.

An opportunity for investors has begun to emerge although many will continue to shun it because of the fear of the reintroduction of controls by a future government or the detailed level of management required. There has been an increase in supply brought about by a number of factors. The tax concessions available under the Business Expansion Scheme (now discontinued) extended to the acquisition of houses for letting subject to limitations on early sales and saw a number of properties being offered to let for terms of up to five years. The subsequent recession increased the supply of houses to let where builders and others failed to find purchasers. Although there is no restriction on rents to be charged on modern lettings, it should be borne in mind that this type of property requires detailed management.

9.3 THE TWO CODES

The two codes have similarities but also contain major differences and are considered separately. Table 9.1 sets out the monetary limits applicable to the respective codes.

9.3.1 Tenancies under the Rent Act 1977

The surviving Rent Act legislation imposes an elaborate system of protection: the different forms it may take will be considered but first the type of occupation affected will be considered briefly. Section 1 of Part I of the 1977 Act provides that 'subject to this part of the Act, a tenancy under which a dwelling house (which may be a house or part of a house) is let as a separate dwelling is a protected tenancy for the purposes of this Act' whilst section 152(1) provides that a tenancy includes a sub-tenancy.

Later legislation has changed the original definition, now requiring one of the following conditions to apply:

Table 9.1 Statutory protection for lettings of residential premises

Description	Class	Appropriate day	Greater London		Elsewhere (England and Wales)	
			Rateable value	Rent	Rateable value	Rent
Protected tenancy	A	on or after 1 April 1973 but before 1 April 1990	£1500		£740	
	B	on or after 22 March 1973 but before 1 April 1973 and had rateable value on appropriate day exceeding	£ 600		£300	
		and on 1 April 1973 exceeding	£1500		£750	
	C	before 22 March 1973 and had rateable value on appropriate day exceeding	£ 400		£200	
		and on 22 March 1973 exceeding	£ 600		£300	
		and on 1 April 1973 exceeding	£1500		£750	
		on or after 1 April 1990		£25 000		£25 000

- The tenancy is one granted to the original tenant before 15 January 1989
- Where a tenant who was a protected or statutory tenant under a tenancy granted before 15 January 1989 and died before that date, the surviving spouse or a resident member is entitled to claim a statutory tenancy by succession on two occasions
- Where a tenant who was a protected or statutory tenant under a tenancy granted before 15 January 1989 and who died on or after that date, the surviving spouse (only) is entitled to claim a statutory tenancy by succession; a qualifying family member obtains an assured tenancy.

Certain transitional rules may apply to the tenancy.

9.3.2 Protected tenancies

The distinguishing feature of a protected tenancy is that it springs from and follows a contractual tenancy. Certain types of tenancy are excepted from the provisions relating to protected tenancies and are set out in sections 4 to 16.

The general proposition is that a letting of a dwelling-house as a separate dwelling is a protected tenancy. The exceptions are given below.

(a) Dwelling-houses outside certain rateable value or rental value limits (section 4)

Where the rateable value at a certain date or dates exceeds a certain figure any tenancy of the dwelling house will not be a protected tenancy. Different ceilings apply according as to what is the 'appropriate day' and as to whether the dwelling is in Greater London or elsewhere.

Where a tenancy is entered into on or after 1 April 1990 it is not protected where the rent exceeds £25 000 per annum.

(b) Tenancies at low rents (section 5)

Where in any tenancy entered into before 1 April 1990 no rent is payable or the rent payable is less than two thirds of the rateable value on the appropriate day. Specified rates apply on and after this date (the provisions as to date and area are included in Table 9.1). 'Rent' will be the total amount payable including any payment in respect of rates: if the rent is reduced to reflect services performed by the tenant, the reduced amount is the figure which will be used to determine whether the tenancy is a protected one. Long tenancies at a low rent gain some protection from Part I Landlord and Tenant Act 1954. The rent shall be determined after the deduction of any amount expressed to be payable in respect of rates, services, repairs, maintenance or insurance.

(c) Dwelling-houses let with other land (section 6)

This exception as set out by section 6 and interpreted in the courts provides that any land or premises let together with a dwelling-house shall be treated as part of the dwelling-house where that use is the dominant purpose of the letting or where it consists of agricultural land exceeding two acres in extent.

(d) Payments for board or attendance (section 7)

The exception does not operate unless the amount of rent which is fairly attributable to attendance, having regard to the value of the attendance to the tenant, forms a substantial part of the whole rent. There is no such requirement in respect of board. But the tenancy may be protected as a restricted contract in respect of contracts entered into before 15 January 1989.

(e) Lettings to students (section 8)

The student must be pursuing, or intending to pursue, a course of study provided by a specified educational institution and having a tenancy granted by that institution or another specified institution or body of persons. The Protected Tenancies (Exceptions) Regulations 1988 (SI No.1988/2236) lists the specific institutions within the scope of this exception.

(f) Holiday lettings (section 9)

It should be noted that this section relates to a tenancy where the purpose is to confer on the tenant the right to occupy the dwelling-house for a holiday without the landlord being in danger of creating a protected tenancy. No definition of 'holiday' is provided.

(g) Agricultural holdings (section 10)

The exception refers to a tenancy of an agricultural holding within the meaning of the Agricultural Holdings Act 1986 where the holding is occupied by the person responsible for the control of the farming (whether as tenant or as servant or agent of the tenant).

(h) Licensed premises (section 11)

The exception applies to tenancies of on-licences, which now enjoy protection under the provisions of Part II of the Landlord and Tenant Act 1954 by virtue of changes enacted by the Landlord and Tenant (Licensed Premises) Act 1990. The exception does not apply to off-licences in cases where there is some living accommodation within the definition of 'dwelling-house'.

(i) Resident landlords (section 12)

A tenancy of a dwelling-house granted on or after 14 August 1974 is not a protected tenancy if:

    (a) the dwelling-house forms part only of a building and except where the dwelling house also forms part of a flat, the building is not a purpose-built block of flats ('flat' is defined as a dwelling-house which 'forms part only of a building and is separated horizontally from another dwelling-house which forms part of the same building') and

   (b) the tenancy was granted by a person who, at the time that he granted it, occupied as his residence another dwelling-house which also forms part of the same flat or same building and

   (c) at all times since the tenancy was granted the interest of the landlord under the tenancy has belonged to a person who, at the time he owned that interest, occupied as his residence another dwelling-house which also formed part of that building.

(Breaks of occupation of up to four years are permitted depending on the circumstances: the most important event is, perhaps, a sale by the landlord in which case a break of more than twenty-eight days will be fatal to continuation of the exception, unless the purchaser notifies the tenant in writing of his intention to occupy as a residence in which case his rights will be preserved for up to six months provided such notice was given within the 28 day period.) The exception does not affect the rights of a tenant previously occupying as a protected or statutory tenant or where a term of years certain is granted to a tenant who occupied earlier but not as a protected tenant.

(j) Landlord's interest belonging to the Crown (section 73)

Certain types of Crown tenant may become regulated tenants. They are tenants of the Duchies of Lancaster and Cornwall and tenants of the Crown Estate Commissioners.

   There are consequential provisions relating to the grounds for possession and to lawful premiums. Similar provisions are made for occupiers of agricultural tied accommodation who may now become statutory tenants. Tenants of the Sovereign's private estates or of a government department are not included in these provisions.

(k) Other exceptions

Other exceptions refer to dwelling-houses where the landlord's interest belongs to a local authority, housing association housing co-operative or charitable housing trust and are therefore beyond the scope of this work.

9.3.3 Security of tenure

A central feature of the legislation is that there is security of tenure whether the tenancy is protected or statutory. No proceedings may be taken unless and until the tenancy has been determined at common law. The landlord must then seek an order of the court: the court shall not make an order for possession unless it considers it reasonable to do so and that either:

1. The court is satisfied that suitable alternative accommodation is available for the tenant or will be available for him when the order takes effect or
2. The circumstances are as specified in any of the Cases in Part I of Schedule 15 of the 1977 Act (with later additions and one repeal there is now a total of nineteen cases).

Where the tenancy has a contractual base, the landlord must first serve a valid notice to quit (valid, that is, in relation between the parties) and which must in any event be in writing, give certain prescribed information and give not less than four weeks' notice.

Where the landlord has a right of re-entry or forfeiture under a lease he can exercise it only by proceedings in court while any person is lawfully residing in the premises or part of the premises. The courts have very wide discretion in deciding what is reasonable. In deciding whether alternative accommodation is suitable the courts have on numerous occasions considered the requirements of the particular rather than the hypothetical tenant. The landlord may offer tenants as alternative accommodation part of the existing demised premises (e.g. the dwelling-house and part only of the garden or part of the living accommodation presently occupied by them) where appropriate. Guidance is given in Part IV of Schedule 15 where it is provided that the accommodation shall be deemed to be suitable if it consists of either:

1. premises which are to be let as a separate dwelling such that they will then be let on a protected tenancy or
2. premises to be let as a separate dwelling on terms which will, in the opinion of the courts, afford to the tenant security of tenure reasonably equivalent to that offered in the case of a protected tenancy.

Where the accommodation is reasonably suitable to the needs of the tenant and his family as regards proximity to place of work and either suitable to the means and needs of the tenant and his family as regards extent and character or similar as regards rental and extent to any dwelling-houses provided in the neighbourhood by any housing authority for persons with similar needs.

Where housing is to be provided by the housing authority for the district a certificate of that authority to the effect that suitable alternative accommodation will be provided by a date specified is to be regarded as conclusive evidence.

9.3.4   Statutory tenancies

The terms, conditions and rules of succession of the statutory tenancy are laid down by sections 2 and 3 and Schedule I of the 1977 Act.

Section 2 provides in part that:

(a) after the termination of a protected tenancy of a dwelling-house the person who, immediately before that termination, was the protected tenant of the dwelling-house shall, if and so long as he occupies the dwelling-house as his residence, be the statutory tenant of it; and

(b) Part I of Schedule I to this Act shall have effect for determining what person (if any) is the statutory tenant of a dwelling-house at any time after the death of a person who, immediately before his death, was either a protected tenant of the dwelling-house or the statutory tenant of it by virtue of paragraph (a) above.

The status of statutory tenant is accorded to former protected tenants and to any person who qualifies as a successor on death of the tenant. Statutory tenants enjoy this personal right only so long as they remain in occupation (though the courts have given a broad interpretation to temporary absence and distinguished that from non-occupation. They have no right to assign or sublet the whole, whatever the original agreement may have provided, though they may sublet part and still retain protection under the Act but perhaps only in respect of the part retained: in such circumstances, however, the subtenant will usually gain protection under the Act.

Section 3 provides in part:

(1) So long as he retains possession, a statutory tenant shall observe and be entitled to the benefit of all the terms and conditions of the original contract of tenancy, so far as they are consistent with the provisions of this Act.

It has been held that an option to purchase the reversion 'at any time' contained in a lease expires when the original tenancy ceases to exist: i.e. it is not carried over to the statutory tenancy.

A statutory tenant is required by section 5 of the Protection from Eviction Act 1977 to give not less than four weeks' notice to quit and longer where so required by the original contract of tenancy. Where no notice was required under that contract, the tenant must give not less than three months' notice. In either case, the notice must be served so as to expire at the end of a rental period. It is provided also that the landlord shall be entitled to access for inspection and to reasonable facilities for executing any repairs. Part II of Schedule I refers to relinquishing tenancies and changing tenants.

Statutory tenants are entitled to receive payment from their landlord as a condition of giving up possession but it is an offence to ask for or receive any payment from any one other than the landlord. Similarly, the purchase of any furniture as a condition of giving up possession is to be at a reasonable price, any excess to be treated as a payment asked for as a condition of giving up possession.

There is provision for a statutory tenant to be replaced by an incoming tenant who is to be deemed the statutory tenant from the transfer date. In order to be effective the landlord must be a party to the agreement which must be in writing. The rights of succession are limited. It is an offence to require a payment for entering into such an agreement, except that outgoing tenants can ask for payment of outgoings referable to any period after the transfer date, their reasonable expenditure on structural alterations or fixtures which they are not entitled to remove, any sum paid by them to their predecessors and, where part of the dwelling is used for business, trade or professional purposes, a reasonable amount in respect of goodwill.

A statutory tenancy may also arise from succession. The spouse of the original statutory tenant becomes the statutory tenant provided that he or she was residing with the tenant immediately before his or her death. Where there is no qualifying spouse, a member of the tenant's family who was residing with them at the time and for a period of six months immediately before their death shall become the statutory tenant and shall remain so as long as they occupy the dwelling-house as their residence. Where there is more than one person so qualified, the statutory tenant shall be the person decided by agreement or in default of agreement by the County Court.

Whether the original tenant's spouse or a member of his or her family succeeds, the successor must be one person only. For the purpose of succession, 'family' has been interpreted widely to include, in addition to the obvious cases, legitimate and adopted children, stepchildren, grandchildren and brothers- and sisters-in-law, the common law husband or wife and cousins and nieces – but the latter two only where justified by the particular circumstances. On the death of the first successor, there are similar provisions for a second succession but not for further successions. A statutory tenancy may be determined by the tenant giving up possession or by his having an order for possession made against him. A statutory tenancy may cease to exist where the tenant no longer occupies the dwelling-house as his residence or where the dwelling-house is destroyed by fire.

9.3.5 Regulated tenancies

A regulated tenancy may be a protected tenancy where the tenant holds a contractual or common law term or a statutory tenancy (including in the latter case a first or second succession) unless it has been released from the provisions of the 1977 Act by an order made by the Secretary of State. A tenancy is not a regulated tenancy if it is a tenancy to which Part II of the Landlord and Tenant Act 1954 applies.

Rent limits may vary according to whether the regulated tenancy is in a contractual or a statutory period and to whether there is a registered rent.

Where during a contractual period of a regulated tenancy a rent is registered (see below), the rent recoverable shall be limited to that rent and where the rent agreed between the parties exceeds that figure, the excess is irrecoverable: should the rent registered be greater than the rent agreed between the parties it will not be possible to recover any sum above the rent agreed prior to the expiration of the contractual period. Where a property is let on a protected tenancy there is no rent limit although it is open to either party to apply for a fair rent to be registered.

Where there is a registered rent, the rent may be increased by a notice specifying a date not more than four weeks prior to the date of the notice and not earlier than the date of registration. Subject to this proviso, the date of registration may be specified as the date for the increase, even though that date falls other than at the beginning of a rental period.

There are provisions for the appointment of Rent Officers in connection with the fixing of fair rents: the Rent Officer for the area must maintain a register of rents available for inspection, a certified copy of any entry therein being available to any person on payment of a prescribed fee. Rent Assessment Committees operate as appeal bodies.

The fair rent is to be arrived at in accordance with the provisions of section 70 of the Rent Act 1977 as amended by the Housing Act 1980, the thrust of the section being to exclude any part of the rental value attributable to scarcity. In view of its importance, the section is quoted in full:

(a) In determining, for the purposes of this Part of this Act, what rent is or would be a fair rent under a regulated tenancy of a dwelling-house, regard shall be had to all the circumstances (other than personal circumstances) and in particular to –

(i) the age, character, locality and state of repair of the dwelling-house, and

(ii) if any furniture is provided for use under the tenancy, the quantity, quality and condition of the furniture.

(b) For the purposes of the determination it shall be assumed that the number of persons seeking to become tenants of similar dwelling-houses in the locality on the terms (other than those relating to rent) of the regulated tenancy is not substantially greater than the number of such dwelling-houses in the locality which are available for letting on such terms.

(c) There shall be disregarded –

(i) any disrepair or other defect attributable to a failure by the tenant under the regulated tenancy or any predecessor in title of his to comply with any terms thereof:

(ii) any improvement carried out, otherwise than in pursuance of the terms of the tenancy, by the tenant under the regulated tenancy or any predecessor in title of his:

((c) and (d) have been repealed)

(iii) if any furniture is provided for use under the regulated tenancy or any predecessor in title of his or, as the case may be, any deterioration in the condition of the furniture due to any ill-treatment by the tenant, any person residing or lodging with him, or any sub-tenant of his.

(d) In this section 'improvement' includes the replacement of any fixtures or fittings.

Where no rent has been registered, there is no restriction on the rent which may be agreed between the landlord and make application to the Rent Officer to fix and register a fair rent.

A tenancy may become a regulated tenancy by conversion: in particular section 64 of the Housing Act 1980 converted all remaining controlled tenancies into regulated tenancies with one exception. The exception to the conversion of a controlled tenancy into a regulated tenancy is where the tenancy comprises mixed business and residential user, in which case the tenancy becomes subject to Part II of the Landlord and Tenant Act 1954 tenancy by conversion unless the business user is merely incidental to the residential user. Subject to this exception, any agreement to an increased rent entered into on or after 28 November 1980 shall be of no effect. Unless the increase is made as a result of a rent fixed by the rent officer, the increase is void and the tenant may recover any excess within one year of payment.

9.4 RESTRICTED CONTRACTS

A restricted contract is one where one person grants to another in consideration of a rent which includes payment for the use of furniture or for services, the right to occupy a dwelling as a residence. Where the dwelling consists of only part of a house it is a restricted contract if the person has exclusive occupation of part, even though other rooms or accommodation in the house are used in common with any other person or persons.

By section 19(8), 'services' includes attendance, the provision of heating or lighting, the supply of hot water and any other privilege or facility connected with the occupancy of a dwelling, other than a privilege or facility requisite for the purpose of access, cold water supply or sanitary accommodation.

In contrast, there is no statutory definition of 'furniture' though it is clear from decided cases that it must consist of more than one or two items and does not include fixtures and fittings in the dwelling nor furniture in the common parts.

Where a tenancy is precluded from being a protected tenancy solely because it was granted by a resident landlord it shall be a restricted contract notwithstanding that the rent may not include payment for the use of furniture or for services.

A contract is not a restricted contract if:

1. the rateable value of the dwelling on the approximate day exceeded certain specified limits (Classes D and E in Table 9.1)
2. it creates a regulated tenancy or
3. under the contract the interest of the lessor belongs to Her Majesty in right of the Crown or to a government department, or is held in trust for Her Majesty for the purpose of a government department or
4. it is a contract for the letting of any premises at a rent which includes payment in respect of board the value of which forms a substantial proportion of the whole or
5. it is a protected occupancy under the Rent (Agriculture) Act 1976 or
6. it creates a tenancy granted by a housing association, housing trust or the Housing Corporation.

A contractual licence may be a restricted contract where board is not provided. A tenancy which is precluded from being a protected tenancy solely because it was granted by a resident landlord shall be treated as a restricted contract even though there is payment for attendance. Similarly, where tenants share certain accommodation with their landlord while enjoying exclusive occupation of other accommodation under a tenancy which does not qualify as a protected tenancy, the tenancy is a restricted contract. Both provisions apply notwithstanding that no payment for the use of furniture or for services is included in the rent. The Secretary of State may by order exclude from these provisions any dwelling where the rateable value exceeds such amount as may be specified: the order may apply generally or only to certain types of dwellings and certain areas of England and Wales.

Arrangements for the review of rents and for their phasing are similar to those described for regulated rents under registered tenancies, though the procedure is different. After 15 January 1989 no tenancy or other contract is capable of being a restricted contract; if the rent under an earlier contract is varied, a new contract is deemed to come into existence which ceases to be a restricted contract except where the variation is by a rent tribunal or by the parties to bring the rent to the level of the registered rent.

9.5 PROTECTED SHORTHOLD TENANCIES

The Act introduces a new type of tenancy to be known as a protected shorthold tenancy. The purpose is to allow landlords to let for a fixed term of between one and five years with an assurance of possession at the end of the period. Tenants enjoy the rent-control provisions which apply to regulated tenancies but have no security of tenure at the end of the term. A protected shorthold tenancy must satisfy the following conditions:

1. It is granted after 28 November 1980.
2. It is granted for a term certain of not less than one and not more than five years and cannot be terminated by the landlord earlier than the expiry of the term except in pursuance of a provision for re-entry or forfeiture for non-payment of rent or breach of any other obligation under the tenancy.
3. Before the grant the landlord has given the tenant a valid notice (to comply with regulations made by the Secretary of State) stating that the tenancy is to be a protected shorthold tenancy.

4. Either there is a regulated rent when the tenancy is granted or a certificate of fair rent has been issued before the grant and the rent payable for any period before a rent is registered does not exceed the rent specified in the certificate and an application is made within 28 days of the beginning of the term and is not withdrawn.
5. The tenant was not a protected or statutory tenant of the dwelling-house immediately before the tenancy was granted.
6. It qualifies in all other respects as a protected tenancy.

Although the tenancy is granted for a term certain, the tenant may terminate the tenancy before the end of the term by giving one month's notice where the original term is two years or less and three months otherwise.

9.6 ASSURED TENANCIES

In an endeavour to encourage investors to build houses to let, the government introduced provisions in the Housing Act 1980 to enable certain approved bodies to let outside the provisions of the Rent Acts but subject to those provisions for renewal of lease and obtaining possession contained in Part II of the Landlord and Tenant Act 1954 and which have hitherto worked well in relation to business premises.

To qualify, construction must have commenced after 8 August 1980 and, prior to the tenant's occupation, no part let other than under an assured tenancy. These provisions have been replaced by more flexible rules under the Housing Act 1988.

9.6.1 Secure tenancies

Secure tenancies are public sector tenancies and are therefore not considered further.

9.7 GROUNDS FOR POSSESSION

There are now nineteen grounds – termed cases – for possession of dwelling-houses let on or subject to protected or statutory tenancies, the first nine of which are discretionary.

(a) Discretionary cases (1 to 10)

The court may order possession under:

Case 1: breach of obligation

Where any rent lawfully due from the tenant has not been paid, or any obligation of the protected or statutory tenancy which arises under this Act, or –

(a) in the case of a protected tenancy, any other obligation of the tenancy, in so far as is consistent with the provisions of Part VII of this Act, or
(b) in the case of a statutory tenancy, any other obligation of the previous protected tenancy which is applicable to the statutory tenancy, has been broken or not performed.

The rent must be lawfully due and unpaid at the date of the commencement of proceedings. Payment of arrears into court after this date will not remove the case from the jurisdiction of the court although it is then less likely to be reasonable to make the order. In any event, the order is likely to be made suspended so long as the rent is paid together with a specified amount off the arrears.

'Any obligation' is wide enough to embrace breaches of implied covenants.

Case 2: Nuisance, etc.

Where the tenant or any person residing or lodging with him or any sub-tenant of his has been guilty of conduct which is a nuisance or annoyance to adjoining occupiers, or has been convicted of using the dwelling-house or allowing the dwelling-house to be used for immoral or illegal purposes.

Nuisance or annoyance together extend to a wide range of possibilities.

Some helpful comments on the second ground of this case were made by Widgery L.J. in *Abrahams* v. *Wilson* (1971).

> If the drugs are on the demised premises merely because the tenant is there and has them in his or her immediate custody . . . then I would say without hesitation that that does not involve a 'using' of the premises in connection with the offence. On the other hand, if the premises are employed as a storage place or hiding place for dangerous drugs, a conviction for possession of such drugs, when the conviction is illuminated by further evidence to show the manner in which the drugs themselves were located, would I think be sufficient to satisfy the section and come within case 2.

> Case 3: deterioration by waste or neglect
>
> Where the condition of the dwelling-house has, in the opinion of the court, deteriorated owing to acts of waste by, or the neglect or default of, the tenant or any person residing or lodging with him or any sub-tenant of his and, in the case of any act of waste by, or the neglect or default of, a person lodging with the tenant or a sub-tenant of his, where the court is satisfied that the tenant has not, before the making of the order in question, taken such steps as he ought reasonably to have taken for the removal of the lodger or sub-tenant, as the case may be.

This is a widely-drawn ground enabling the landlord to obtain a remedy where the property is suffering deterioration.

Case 4: deterioration of furniture by ill-treatment

Here the condition of any furniture provided for use under the tenancy has, in the opinion of the court, deteriorated owing to ill-treatment by the tenant or any person residing or lodging with him or any sub-tenant of his and, in the case of any ill-treatment by a person lodging with the tenant or a sub-tenant of his, where the court is satisfied that the tenant has not, before the making of the order in question, taken such steps as he ought reasonably to have taken for the removal of the lodger or sub-tenant, as the case may be.

Cases 3 and 4 may be seen together and reflect the common law obligations of a tenant to behave in a tenant-like fashion.

Case 5: tenant's notice to quit

Where the tenant has given notice to quit and, in consequence of that notice, the landlord has contracted to sell or let the dwelling-house or has taken any other steps as the result of which he would, in the opinion of the court, be seriously prejudiced if he could not obtain possession.

There must be a valid notice to quit (as to which see above).

The term 'seriously prejudiced' suggests that the landlord had committed himself contractually for the sale or letting of the premises. Perhaps it would extend to a case where the landlord had committed himself to expenditure in anticipation of selling the property with vacant possession.

Case 6: assignment or subletting

This case applies even though the contractual tenancy contains no covenant prohibiting assignment and subletting. A statutory tenant has no power of assignment.

Case 7: repealed

Case 8: dwelling required for landlord's employee

Where the dwelling-house is reasonably required by the landlord for occupation as a residence for some person engaged in his whole-time employment, or in the whole time employment of some tenant from him or with whom, conditional on housing being provided, a contract for such employment has been entered into, and the tenant was in the employment of the landlord or a former landlord, and the dwelling-house was let to him in consequence of that employment and he has ceased to be in that employment.

The court may order payment of compensation where it appears to the court that an order for possession was obtained by misrepresentation or concealment of material facts. The amount of compensation is to be '. . . such sum as appears sufficient as compensation for damage or loss sustained by that tenant as a result of the order'. There are separate provisions relating to dwelling houses occupied by agricultural workers (see cases 16 to 18, section 9.3.7 and section 9.7).

Case 9: dwelling required by landlord for own occupation

Where the dwelling-house is reasonably required by the landlord for occupation as a residence for –

(a) himself, or
(b) any son or daughter of his over 18 years of age, or
(c) his father or mother, or
(d) if the dwelling-house is let on or subject to a regulated tenancy, the father or mother of the landlord's wife or husband.

and the landlord did not become landlord by purchasing the dwelling-house or any interest therein after 23 March 1965, or 8 March 1973 in respect of tenancies which became regulated on that date or 24 May 1974 in respect of furnished tenancies.

Schedule 15 provides in Part III:

> A court shall not make an order for possession of a dwelling-house by reason only that the circumstances of the case fall within Case 9 in Part I of this Schedule if the court is satisfied that, having regard to all the circumstances of the case, including the question whether other accommodation is available for the landlord or the tenant, greater hardship would be caused by granting the order than by refusing to grant it.

As will be seen, the parties for whom the dwelling-house may be required are limited and the relationships are more restricted than those in the case of statutory succession. Where the dwelling-house was purchased with vacant possession the provisions restricting use of this case do not apply: the reference is specifically to becoming a landlord by purchase.

Case 10: subletting of part at an excessive rent

Where the court is satisfied that the rent charged by the tenant –

(a) for any sublet part of the dwelling-house which is a dwelling-house let on a protected tenancy or subject to a statutory tenancy is or was in excess of the maximum rent for the time being recoverable for that part, having regard to Part III of this Act, or

(b) for any sublet part of the dwelling-house which is subject to a restricted contract is or was in excess of the maximum (if any) which it is lawful for the lessor, within the meaning of Part v. of this Act to require or receive having regard to the provisions of that Part.

(b) Mandatory cases (11 to 20)

The mandatory grounds for possession are set out below. In particular it is necessary that not later than the relevant date the landlord gave notice in writing to the tenant that possession might be recovered under the specified case. The 'relevant date' for mandatory cases is the date of commencement of the regulated tenancy except where:

- a tenancy became a regulated tenancy by virtue of section 73 of the Housing Act 1980 (dwellings owned by the Crown Estates or by a government department) in which case the relevant date is 8 February 1981 or

- a regulated furnished tenancy was created before 14 August 1974 when the relevant date is 13 February 1975 or
- a tenancy created before 22 March 1973 became a regulated tenancy by virtue of the Counter-Inflation Act 1973 when the relevant date is 22 September 1973
- a protected tenancy was created before 8 December 1965 when the relevant date is 7 June 1966.

Case 11: the landlord is an owner–occupier

Where a person (the owner occupier) who occupies the dwelling-house on a regulated tenancy had, at any time before the letting, occupied it as his residence and –

(a) not later than the relevant date the landlord gave notice in writing to the tenant that possession might be recovered under this Case, and

(b) the dwelling-house has not, since one of three specified dates been let on a protected tenancy not subject to a case 11 notice and

(c) the court is of the opinion that one of the conditions set out below is satisfied:

(The court has discretion to dispense with the requirements of either or both paragraphs (a) and (b) above where it is of the opinion that it is just and equitable to do so.)

(a) the dwelling-house is required as a residence for the owner or any member of his family who resided with the owner when he last occupied the dwelling-house as a residence; or

(b) the owner has died and the dwelling-house is required as a residence for a member of his family who was residing with him at the time of his death; or

(c) the owner has died and the dwelling-house is required by a successor in title as his residence or for the purpose of disposing of it with vacant possession; or

(d) the dwelling-house is subject to a mortgage, made by deed and granted before the tenancy, and the mortgagee is entitled to exercise a power of sale conferred on him by the mortgage or by section 101 of the Law of Property Act 1925; and requires the dwelling-house for the purpose of disposing of it with vacant possession in exercise of that power; or

(e) the dwelling-house is not reasonably suitable to the needs of the owner, having regard to his place of work, and he requires it for the purpose of disposing of it with vacant possession and of using the proceeds of that disposal in acquiring, as his residence, a dwelling-house which is more suitable to those needs.

Case 12: retirement home

Where the landlord (in this Case referred to as 'the owner') intends to occupy the dwelling-house as his residence at such time as he might retire from regular employment and has let it on a regulated tenancy before he has so retired and –

(a) not later than the relevant date the landlord gave notice in writing to the tenant that possession might be recovered under this Case; and

(b) the dwelling-house has not, since 14 August 1974, been let by the owner on a protected tenancy with respect to which the condition mentioned in paragraph (a) above was not satisfied; and the owner has retired and requires the dwelling-house as a residence for himself or

(c) the court is of the opinion that conditions (b) to (d) of Case 11 apply.

If the court is of the opinion that, notwithstanding that the condition in paragraph (a) or (b) above is not complied with, it is just and equitable to make an order for possession of the dwelling-house, the court may dispense with the requirements of either or both of those paragraphs, as the case may require.

The practical effect of this case is to allow purchase and letting of premises intended for occupation in retirement: the case is of importance in particular to occupiers of 'tied' accommodation who wish to acquire a home some time before it will be required and to obtain an income from it in the meanwhile.

Case 13: out-of-season lettings

Where the dwelling-house is let under a tenancy for a term of years certain not exceeding 8 months and –

(a) not later than the relevant date the landlord gave notice in writing to the tenant that possession might be recovered under this Case; and

(b) the dwelling-house was, at some time within the period of 12 months ending on the relevant date, occupied under a right to occupy it for a holiday.

For the purposes of this Case a tenancy shall be treated as being for a term of years certain notwithstanding that it is liable to determination by re-entry or on the happening of any event other than the giving of notice by the landlord to determine the term.

This case is intended to cover out-of-season lettings of holiday homes where the dwelling-house had been occupied (not necessarily let) for a holiday within a previous specified period.

Case 14: short tenancies of student accommodation

Where the dwelling-house is let under a tenancy for a term of years certain not exceeding 12 months and –

(a) not later than the relevant date the landlord gave notice in writing to the tenant that possession might be recovered under this Case; and

(b) at some time within the period of 12 months ending on

the relevant date, the dwelling-house was subject to such a tenancy.

For the purposes of this Case a tenancy shall be treated as being for a term of years certain notwithstanding that it is liable to determination by re-entry or on the happening of any event other than the giving of notice by the landlord to determine the term.

This case refers to 'vacation lettings' and enables specified educational institutions to let accommodation for short periods to non-students without granting security of tenure.

Case 15: dwelling-house for minister of religion

Here the dwelling-house is held for the purpose of being available for occupation by a minister of religion as a residence from which to perform the duties of his office and –

(a) not later than the relevant date the tenant was given notice in writing that possession might be recovered under this Case, and
(b) the court is satisfied that the dwelling-house is required for occupation by a minister of religion as such a residence.

Case 16: dwelling house for agricultural employee

Where the dwelling-house was at any time occupied by a person under the terms of his employment as a person employed in agriculture, and

(a) the tenant neither is nor at any time was so employed by the landlord and is not the widow of a person who was so employed, and
(b) not later than the relevant date, the tenant was given notice in writing that possession might be recovered under this Case, and

(c) the court is satisfied that the dwelling-house is required for occupation by a person employed, or to be employed, by the landlord in agriculture.

For the purposes of this Case 'employed', 'employment' and 'agriculture' have the same meanings as in the Agricultural Wages Act 1948.

Case 17

Where proposals for amalgamation, approved for the purposes of a scheme under section 26 of the Agriculture Act 1967, have been carried out and, at the time when the proposals were submitted, the dwelling-house was occupied by a person responsible (whether as owner, tenant, or servant or agent of another) for the control of the farming of any part of the land comprised in the amalgamation and

(a) after the carrying out of the proposals, the dwelling-house was let on a regulated tenancy otherwise than to, or to the widow of, either a person ceasing to be so responsible as part of the amalgamation or a person who is, or at any time was, employed by the landlord in agriculture, and
(b) not later than the relevant date the tenant was given notice in writing that possession might be recovered under this Case, and
(c) the court is satisfied that the dwelling-house is required for occupation by a person employed, or to be employed, by the landlord in agriculture, and
(d) the proceedings for possession are commenced by the landlord at any time during the period of 5 years beginning with the date on which the proposals for the amalgamation were approved or, if occupation of the dwelling-house after the amalgamation continued in, or was first taken by, a person ceasing to be responsible as mentioned in paragraph (a) above or his widow, during a period expiring 3 years after the date on which the dwelling-house next became unoccupied.

For the purposes of this Case 'employed' and 'agriculture' have the same meanings as in the Agricultural Wages Act 1948 and 'amalgamation' has the same meaning as in Part II of the Agriculture Act 1967.

Case 18

Where –

(a) the last occupier of the dwelling-house before the relevant date was a person, or the widow of a person, who was at some time during his occupation responsible (whether as owner, tenant, or servant or agent of another) for the control of the farming land which formed, together with the dwelling-house, an agricultural unit within the meaning of the Agriculture Act 1947, and

(b) the tenant is neither –

(i) a person, or the widow of a person, who is or has at any time been responsible for the control of the farming of any part of the said land, nor

(ii) a person, or the widow of a person, who is or at any time was employed by the landlord in agriculture, and

(c) the creation of the tenancy was not preceded by the carrying out in connection with any of the said land of an amalgamation approved for the purposes of a scheme under section 26 of the Agriculture Act 1967, and

(d) not later than the relevant date the tenant was given notice in writing that possession might be recovered under this Case, and

(e) the court is satisfied that the dwelling-house is required for occupation either by a person responsible or to be responsible (whether as owner, tenant, or servant or agent of another) for the control of the farming of any part of the said land or by a person employed or to be employed by the landlord in agriculture, and

(f) in a case where the relevant date was before 9 August 1972, the proceedings for possession are commenced by the landlord before the expiry of 5 years from the date on which the occupier referred to in paragraph (a) above went out of occupation.

For the purposes of this Case 'employed' and 'agriculture' have the same meanings as in the Agricultural Wages Act 1948 and 'amalgamation' has the same meaning as in Part II of the Agriculture Act 1967.

Case 19: protected shorthold tenancies

Where the dwelling-house was let under a protected shorthold tenancy (or is treated under section 55 of the Housing Act 1980 as having been so let) and –

(a) there has been no grant of a further tenancy of the dwelling-house since the end of the protected shorthold tenancy or,
(b) if there has been such a grant, it was to a person who was not, immediately before the grant, in possession as a protected or statutory tenant; and
(c) the proceedings for possession were commenced after appropriate notice by the landlord to the tenant and not later than 3 months after the expiry of the notice.

A notice is appropriate for this Case if it –

(a) is in writing and states that proceedings for possession under this Case may be brought after its expiry; and
(b) expires not earlier than 3 months after it is served nor, if, when it is served, the tenancy is a periodic tenancy, before that periodic tenancy could be brought to an end by a notice to quit served by the landlord on the same day;
(c) is served –

(i) in the period of three months immediately preceding the date on which the protected shorthold tenancy comes to an end; or

(ii) If that date has passed, in the period of three months immediately preceding any anniversary of that date; and

(iii) in a case where a previous notice has been served by the landlord on the tenant in respect of the dwellinghouse, and that notice was an appropriate notice, it is served not earlier than 3 months after the expiry of the previous notice.

A relaxation of these provisions is introduced by section 55(2) of the Housing Act 1980 which provides –

If, in proceedings for possession under Case 19 set out above, the court is of opinion that, notwithstanding that the condition of paragraph (b) or (c) of section 52(1) above is not satisfied, it is just and equitable to make an order for possession, it may treat the tenancy under which the dwelling-house was let as a protected shorthold tenancy.

Case 20: lettings by servicemen

Where the dwelling-house was let by a person (in this Case referred to as 'the owner') at any time after the commencement of section 67 of the Housing Act 1980 (28 November 1980):

(a) at the time when the owner acquired the dwelling-house he was a member of the regular armed forces of the Crown;

(b) at the relevant date the owner was a member of the regular armed forces of the Crown;

(c) not later than the relevant date the owner gave notice in writing to the tenant that possession might be recovered under this Case;

(d) the dwelling-house has not, since the commencement of section 67 of the Act of 1980 been let by the owner on a protected tenancy with respect to which the condition mentioned in paragraph (c) above was not satisfied; and

(e) the court is of the opinion that –

(i) the dwelling-house is required as a residence for the owner; or

(ii) of the conditions set out in Part v. of the schedule one of those in paragraphs (c) to (f) is satisfied.

If the court is of the opinion that, notwithstanding that the condition in paragraph (c) or (d) above is not complied with, it is just and equitable to make an order for possession of the dwelling-house, the court may dispense with the requirements of either or both of these paragraphs, as the case may require.

It will be noted that there is no requirement of prior occupation here as is necessary to satisfy Case 11.

For the purposes of this Case 'regular armed forces of the Crown' has the same meaning as in section 1 of the House of Commons Disqualification Act 1975.

9.8 MISCELLANEOUS MATTERS

A number of other provisions apply in particular circumstances.

9.8.1 Sublettings

The protection afforded to tenants is personal to them and will be lost unless the tenant is in possession of at least part of the premises and that part is protected under the Acts.

Occupation is an essential element in the status of a statutory tenant.

Section 23 of the Rent Act 1977 provides that a tenant shall not lose protection solely because a subletting of any part of the premises includes the sharing of accommodation or the provision of board or attendance.

Where a court makes an order for possession against a protected or statutory tenant under one of the cases in the 1977 Act the order does not extend to a subtenant to whom the dwelling-house or any part of it had been lawfully sublet before the commencement of proceedings for possession. The subtenant becomes the tenant of the landlord on the same terms as previously enjoyed by the tenant: where only part of the premises is occupied by the subtenant the rent payable under the superior tenancy is to be apportioned.

Where the subtenancy includes provision by the immediate landlord of furniture or services the superior landlord has an opportunity to avoid such parts of the agreement by notice within six weeks of the termination of the statutorily protected tenancy.

Tenants are required to give to their landlord details of any subletting, including the rent charged, within 14 days of the subletting, except where the terms are the same as a previously notified subletting of that part.

Subletting itself offers two discretionary grounds on which the court may grant possession:

1. Case 6 where, without the consent of the landlord, the tenant has assigned or sublet the whole of the dwelling-house or sublet part of the dwelling-house, the remainder being already sublet;
2. Case 10 where the court is satisfied that the rent charged by the tenant for any sublet part of the dwelling-house is in excess of the maximum rent for the time being recoverable for that part.

Finally, in respect of assured tenancies the landlord may oppose a new tenancy on what may be termed economic grounds, where the current tenancy was created by the subletting of part only of the property and the landlord can show that the aggregate of the rents reasonably obtainable on separate lettings would in total be substantially less than the rent reasonably obtainable on a letting of that property as a whole.

9.8.2 Rights of a spouse

The Matrimonial Homes Act 1983 makes detailed provisions for the non-tenant spouse in occupation of the dwelling-house to be protected from eviction and where not in occupation a right with leave of the court to enter and occupy. Possession by the non-tenant spouse is to be treated as possession by the tenant spouse, all as long as the marriage subsists. There are further provisions whereby the court is empowered as regards parties to divorce, nullity or judicial separation proceedings to transfer a statutory tenancy to the non-tenant party.

9.8.3 Premiums

For the purposes of the Rent Acts a premium is defined as including:

(a) any fine or other like sum;
(b) any other pecuniary consideration in addition to rent; and
(c) any sum paid by way of a deposit, other than one which does not exceed one-sixth of the annual rent and is reasonable in relation to the potential liability in respect of which it is paid.

It is also unlawful to require or to accept a premium, as defined above, as a condition of the grant, renewal or continuance of a protected tenancy. In addition to imposing a fine, the court may order the amount of the premium to be repaid.

Where the purchase of any furniture is required as a condition of the grant, renewal, continuance or assignment of a protected tenancy the amount, if any, by which the amount exceeds a reasonable price for the furniture shall be treated as if it were a premium.

Where a protected regulated tenancy is to be granted, continued or renewed any requirement that rent shall be payable before the beginning of the rental period or earlier than six months before the end of the rental period where that period is more than six months, shall be void and the rent irrecoverable. Where such payments have been made they may be recovered for up to two years after payment by deduction from rent or from the landlord or their personal representatives. These payments are regarded as rent in advance and avoidable in the circumstances given. Additionally, illegal premiums paid may be recovered from the person to whom they were paid (not necessarily the landlord).

There are separate provisions in respect of premiums in relation to certain long tenancies.

9.8.4 Service charges

The provision of services by the landlord, reimbursed by the payment of a service charge by the tenant or tenants, is an area which has offered loopholes to the landlord in the past. There have been several attempts to provide effective control; the legislative provisions are now contained in the Landlord and Tenant Acts 1985 and 1987 and are considered later.

9.8.5 Rent books

There is a statutory requirement on the landlord of premises let as a residence in consideration of a rent payable weekly to provide a rent book or other similar document.

The provision does not apply to any premises where the rent includes a payment in respect of board and the value of that board to the tenant forms a substantial proportion of the whole rent.

The rent book must give the name and address of the landlord together with other prescribed information with penalties for failing to do so. The requirements are set out in the Rent Book (Forms of Notice) Regulations 1982 (SI 1982/1474 as amended by SIs 1988/2198, 1990/1067 and 1993/656). The requirement applies only where the rent is payable weekly, whatever the length of the tenancy.

9.8.6   Protection from unlawful re-entry or eviction

The Protection from Eviction Act 1977 provides protection from unlawful eviction and harassment. The Act is concerned with unlawful eviction and harassment of residential occupiers, whether they occupy under a contract or by virtue of any enactment or rule of law allowing them to remain in possession or restricting the right of any other person to recover possession.

Any person who unlawfully deprives or attempts to deprive occupiers of their occupation or engages in acts of harassment calculated to interfere with their peace or comfort or persistently withdraws or withholds services reasonably required with intent to cause them to give up occupation or to refrain from exercising any right in respect of the premises or part thereof shall be guilty of a criminal offence for which penalties are laid down. There may also be a claim in civil proceedings.

9.8.7   Eviction only by due process of law

The owner may not enforce against the occupiers his right to recover possession of the premises where a tenancy which is not a statutorily protected or an excluded tenancy comes to an end, where the occupier continues in occupation of the premises or part of them, otherwise than by proceedings in the court. This provision also applies where the owner's right to recover possession arises on the death of the tenant under a statutory tenancy and to a licence created under the Rent Act 1977. The provisions do not apply to any excluded tenancy or licence.

9.8.8   Damages

Reinstatement of the occupier before the proceedings are finally disposed of or where the court orders reinstatement at the request of the occupier avoids liability on the part of the landlord or person acting on their behalf. Any damages may be mitigated by the conduct of the occupiers or by their unreasonable refusal to accept reinstatement or alternative accommodation. Damages are based on the difference in value immediately before the residential occupier ceased to occupy the premises between:

1. the value of the leasehold interest subject to the occupier's continuing rights of occupation and
2. the value of the leasehold interest without that right

on the assumption that the landlord is selling his interest in the open market to a willing buyer, excluding the occupier or a member of his family and that any demolition or redevelopment is unlawful.

9.8.9 Restriction on re-entry

Where the premises are let on a lease subject to a right of re-entry or forfeiture there must nevertheless be proceedings in court to enforce the right while any person is lawfully residing on the premises or part of them. This section is specifically provided to be binding on the Crown.

9.8.10 Validity of notices to quit

No notice to quit any premises let as a dwelling by a landlord or a tenant shall be valid unless it is in writing, contains prescribed information and is given not less than four weeks before the date on which it is to take effect. The same rule applies to periodic licences except where terminable with the licensee's employment but not to excluded tenancies or licences.

9.9 DISTRESS

Distress is an ancient self-help remedy whereby a landlord, either personally or through a bailiff, seizes goods belonging to the tenant in order to raise sufficient money on sale to discharge arrears of rent. So far as protected, statutory and assured tenancies are concerned, no distress shall be levied except by leave of the county court which has wide discretionary powers to adjourn proceedings or to stay, suspend or postpone any order.

The Law Commission has proposed the abolition of this remedy (Law Com No. 194, 1991).

9.10 TIED ACCOMMODATION

A landlord–tenant relationship may arise by virtue of the fact that the occupier of the dwelling-house is employed by the owner. Occupiers may be in possession by virtue of a tenancy agreement or of a licence: the accommodation may or may not be essential to the proper execution of their employment. The precise facts of occupation will determine the extent and form of the protection available.

Where there is a service tenancy (as opposed to a licence) the Rent Acts will apply in which case there is a discretionary ground for possession available to the landlord in case 8. Where occupation is not within the provisions of the Rent Acts, it is nevertheless provided that any person who, under the terms of his employment, had exclusive possession of any premises other than as a tenant shall be deemed to have been a tenant in relation to the procedures for regaining possession by the landlord.

9.11   PROTECTION FOR AGRICULTURAL WORKERS

There are special provisions intended to protect agricultural workers occupying dwelling houses either as tenants or as licensees.

Where occupation is by virtue of an agreement entered into before 15 January 1989, protection will be derived from the Rent (Agriculture) Act 1976. Where occupation arose under an agreement entered into on or after that date it will be subject to the Housing Act 1988 as an assured agricultural occupancy.

9.11.1   Occupations prior to 15 January 1989

Special provisions are necessary in relation to houses occupied by agricultural workers because they would only rarely be protected by the Rent Acts. For example, many houses are occupied rent free or at a nominal rent: sometimes some meals are provided also. Occupation comes about contractually and is referred to as a protected occupancy: the Act does not come into operation unless and until the contract is terminated, when it is converted into a statutory tenancy (but of a type created especially for the purposes of the Act and not to be confused with a statutory tenancy under the Rent Acts).

Where the tenancy qualifies under the Rent Acts the tenant is entitled to the protection afforded by those Acts. The rights of the statutory tenant under the 1976 Act are akin to those provided by the Rent Acts although some matters of detail differ: principally the grounds on which possession may be ordered, succession limited to one occasion and protection where the accommodation is shared with anyone other than the landlord or the employer.

The terms of the new statutory tenancy are implied by the Act. They include a provision that the rent payable must not exceed the registered rent if there is one or otherwise an amount based on the rateable value on 31 March 1990. If the rent is fixed by agreement between the parties and is then registered, that rent becomes the rent limit. The rent is payable weekly in arrears. The landlord will have a statutory obligation to repair under the Landlord and Tenant Act 1985 and must not prevent reasonable access or discontinue the provision of necessary services previously provided: the tenant must use the premises in a tenant-like manner, allow the landlord access to carry out repairs, use the premises as a private dwelling-house only and not assign or sublet the whole or any part.

The Act provides ten cases under which the County Court may and two cases under which the court must, make an order for possession on the application of the landlord. The most important case is the discretionary one which enables the landlord to serve notice on the occupier where they have shown to the satisfaction of the housing authority that the dwelling-house is required to house an employee or prospective employee in agriculture, that they are unable to provide suitable alternative accommodation, that such provision is in the interests of efficient agriculture and the housing authority has offered suitable alternative accommodation.

The housing authority must use its best endeavours to provide suitable alternative accommodation where a landlord applies in writing and is able to show that they are unable to provide suitable alternative accommodation and that it is in the interests of efficient agriculture that such accommodation should be provided by the authority.

Advisory Committees may be approached by any of the parties for their advice which must be taken into account by the authority when determining whether it has an obligation to rehouse the occupier. The housing authority must make its decision having regard to the advice given by the Committee, the urgency of the case, other needs for the accommodation available and the resources available to it. Any breach of the statutory duty, such as it is, may be pursued by an action for damages against the housing authority.

Finally, three mandatory cases are provided under the Rent Act 1977 (Cases 16, 17 and 18) where possession may be obtained for purposes connected with agriculture (see Section 9.7(b)).

9.12 MULTIPLE OCCUPATION

Where a house is occupied by more than one family it is in multiple occupation and subject to special provisions relating to repair and maintenance, overcrowding, provision of standard amenities and means of escape from fire. A local authority has power to make a management order appointing a manager to be responsible for compliance with the regulations, a copy of which must be displayed in the house: this may be followed by a control order under which the local authority takes control.

Special grants are available towards the provision of standard amenities.

9.13 REPAIRS, IMPROVEMENTS AND GRANTS

Legislators have attempted to ensure the proper maintenance and improvement of the housing stock by a combination of sanctions and incentives. Thus, local authorities have wide powers to deal with disrepair while various enactments have offered some advantage to the landlord where property is in proper repair. More recently, grants have been made available for improvement.

9.13.1 Landlord's implied obligations

The common law position relating to implied repairing obligations is not altogether certain. In a lease of a house in course of erection, there is an implied warranty that it will be built of proper materials in a competent manner so as to be fit for human habitation on completion. In the absence of express terms to the contrary, there is an implied condition in relation to furnished (but not unfurnished) premises that they will be in habitable condition when let. Similarly, where a building is let to a number of tenants without any express duty on any of them, it is implied that the landlord is responsible for the repair of the common parts.

It has been argued that the landlord's duty to repair does not arise until the want or repair has been notified to him although it is suggested that where the landlord expressly or impliedly reserves a right of entry to inspect the condition of the premises, he will be responsible for the work together with any consequences of his failure to do so, in any instance where he was, or should have been, aware of the disrepair.

In the case of most dwelling-houses, the position is now controlled by statute. The Landlord and Tenant Act 1985 governs the repairing obligations of landlords of dwelling-houses let on lease for less than three years. The act implies a condition that the house is at the commencement of the tenancy fit for human habitation and an undertaking that it will be so maintained by the landlord during the tenancy but only in the case of certain houses let at very low rents.

9.13.2 Landlord's statutory obligations

Landlords have repairing obligations imposed on them where the lease is granted on or after 24 October 1961 for a term less than seven years and except on leases granted on or after 3 October 1980 to a local authority and certain other public sector bodies. Where a lease for a longer term is determinable at the landlord's option within that period it is also subject to the provisions of the act; an option on the part of the tenant enabling the lease to run for more than seven years takes the lease out of the Act. The landlord's responsibilities implied under the Act are to:

- keep in repair the structure and exterior of the house including drains, gutters and external pipes and
- to keep in repair and proper working order the installation for the supply of water, gas and electricity and for sanitation including basins, sinks, baths and sanitary conveniences but not other fixtures, fittings and appliances for making use of the supply of water, gas or electricity and installations for space heating and heating water.

9.13.3 Tenants' implied obligation

Where tenants have not entered into express covenants to repair, they have an implied obligation to use the premises in a tenant-like manner and not to commit waste. The implied obligation will exclude fair wear and tear and therefore imposes a very low standard of maintenance.

9.14 LICENCES

A licence gives a right to occupy but conveys no legal estate or interest and may not give exclusive possession. It is clear that without exclusive occupation a tenancy cannot exist but it does not follow that exclusive possession will necessarily be conclusive in defeating a claim to the existence of a licence.

9.14.1 Licence versus tenancy

The importance of the distinction in the occupation of residential accommodation is that the Rent Acts do not in general apply to licences (although see the provisions relating to restricted contracts, page 221). While it is not possible to contract out of the provisions of the Acts, it may be possible to arrange the transaction in such a way that it falls outside the legislation. The transaction must be a genuine one and not a sham merely to avoid the Acts. Denning M.R. described the test in this way:

> All the circumstances have to be worked out. Eventually the answer depends on the nature and quality of the occupancy. Was it intended that the occupier should have a stake in the room or did he have only permission for himself personally to occupy the room, whether under a contract or not, in which case he is a licensee.
>
> *Marchant* v. *Charters* (1977)

The Protection from Eviction Act 1977 specifically gives protection to, inter alia, 'a licensee, whether or not he is a lessee under a restricted contract and whether or not he has exclusive possession of any part of the premises'. But a licence is not capable of being an assured tenancy.

9.14.2 Licences generally

The House of Lords considered the question of licences in *Street* v. *Mountford* (1985) where an occupier was granted exclusive possession by what was contended by the owner to be a licence in accordance with the intention of the parties. The immediate difficulty was that exclusive possession coupled with a periodic payment would normally point to a tenancy. In determining that the particular arrangement constituted a tenancy, Lord Templeman said 'if exclusive possession at a rent for a term does not constitute a tenancy, then the distinction between a contractual tenancy and a contractual licence becomes wholly unidentifiable'.

The courts will be understandably anxious to ensure that legislation put in place to protect vulnerable sections of society cannot easily be avoided.

9.15 TENANCIES UNDER THE HOUSING ACT 1988

The Housing Act 1988 introduced assured tenancies which provide some security of tenure to assured tenants (except assured shorthold tenants) but no longer offer rent control to new tenancies.

As before, the landlord cannot lawfully evict the tenant without an order of the county court. In the case of an assured tenancy, the landlord will be required to show to the satisfaction of the court that he can satisfy one of the mandatory or discretionary grounds provided by the act. An assured shorthold tenant has no security of tenure once the tenancy has been lawfully determined.

To protect the tenant, an assured tenancy cannot be granted to any person who, immediately before the grant, was a protected or statutory tenant under the 1977 Act. The new provisions enable landlords to grant tenancies at market rents and in respect of lettings since 15 January 1989 there is no requirement for the approval of the landlord by the Secretary of State.

9.15.1 What is an assured tenancy?

By section 1 of the Housing Act 1988 an assured tenancy is a tenancy under which a dwelling house is let as a separate dwelling subject to the continuation of two conditions:

1. the tenant or each of the joint tenants is an individual and
2. the tenant or at least one of the joint tenants occupies the dwelling-house as their only or principal home.

A dwelling-house may be a house or part of a house. The tenancy may be periodic or fixed term; neither affects its ability to be an assured tenancy.

9.15.2 What is not an assured tenancy?

The Act contains a schedule of tenancies which cannot be assured or assured shorthold:

1. tenancies entered into before the commencement of the Housing Act 1988;
2. tenancies of dwelling houses with high rental or rateable values;
3. tenancies at a low rent;
4. business tenancies;
5. tenancies of licensed premises;
6. tenancies of agricultural land;
7. tenancies of agricultural holdings;
8. student lettings;
9. holiday lettings;
10. tenancies granted by resident landlords;
11. Crown tenancies;
12. tenancies granted by local authorities and other bodies;
13. transitional cases.

9.15.3 Regaining possession of assured tenancies

The tenant under an assured fixed-term tenancy is entitled to remain in possession under a periodic assured tenancy unless the landlord obtains an order of court. The only circumstances under which the court can make an order for possession when the dwelling-house is let on an assured, fixed-term tenancy is where mandatory grounds 2 or 8 or any of the discretionary grounds (except 9 and 16) apply, provided that the agreement itself makes provision for termination on the ground used. Where the tenancy is an assured periodic one, the landlord must establish one or more of the statutory grounds provided by the Act. The landlord must have served a notice to quit in accordance with the Act (Assured Tenancies and Agricultural Occupancies (Forms) Regulations 1988 SI 1988/2203).

In the case of assured shorthold tenancies, the Act sets out a means of regaining possession.

The grounds number 16 in total, eight mandatory and the remainder discretionary. Some are new; others will be recognised as being similar to those under the Rent Act 1977 although sometimes with important changes. The grounds are as follow:

Mandatory

1. required for owner occupation of which notice was given not later than the beginning of the tenancy;
2. repossession by the mortgagee, subject to a notice having previously been given by the landlord under ground 1 or by the mortgagee;
3. out of season lettings where notice was given as before that possession might be recovered on this ground;
4. short tenancies of student accommodation where notice was given;
5. a house required for the housing of a Minister of religion subject to a prior notice requirement;
6. redevelopment by the landlord where the works cannot be carried out without gaining possession;
7. recovery of possession following the death of an assured periodic tenant (and assuming no succession rights);
8. serious rent arrears;

Discretionary

9. suitable alternative accommodation is or will be available to the tenant;

10. rent lawfully due from the tenant remains unpaid;
11. persistent delay in paying rent on the part of the tenant;
12. breach of any obligation of the tenancy (other than as to rent);
13. the dwelling-house or any of its common parts has deteriorated owing to acts of waste by or through the default of the tenant;
14. the tenant or any other occupier is guilty of causing a nuisance or annoyance to adjoining occupiers;
15. the condition of any furniture supplied has deteriorated owing to ill treatment by the tenant or any other occupier;
16. the dwelling house was let to the tenant in consequence of his employment by the present or a previous landlord and his employment has now ceased.

The division between mandatory and discretionary cases is not always as in the earlier code. For these reasons, the grounds should be seen as quite separate.

9.16 SAFEGUARDS PROVIDED TO TENANTS OF FLATS

The Landlord and Tenant Act 1987 established a right of first refusal for qualifying tenants of residential flats, gave powers of compulsory purchase for premises suffering from disrepair as a result of the landlord's breach of covenant and strengthened the provisions of the Landlord and Tenant Act 1985 in the matter of service charges.

9.16.1 First refusal on disposal

To qualify, the building must consist of two or more flats where the number of flats held by qualifying tenants must exceed half the total number of flats. The legislation does not apply where the interest of the landlord is exempt, where the landlord is resident or where the internal floor area of any non-residential parts exceeds 50% of the area. Disposals may not be relevant disposals (e.g. mortgage, will, option, gift within family, etc.). Tenants do not qualify where they are tenants not only of the flat under discussion but also of at least two other flats in the premises or where they hold on a protected shorthold or assured tenancy, as business tenants or as tenants by virtue of their employment.

Where the landlord proposes to make a disposal where the Act applies he must serve a formal offer notice on all or at least 90% of the qualifying tenants or all but one where there are fewer than ten qualifying tenants.

Information must include details of the estate or interest and the consideration sought and give an initial two months' period for acceptance
with a further two months for notification of the name of the person who
is to acquire the landlord's interest. An acceptance notice must be from a
majority of qualifying tenants to be valid; the landlord is then precluded
from disposing of the interest unless and until the end of the second
period or, if a person is nominated, for that period and a further three
months. Where no one is nominated following an acceptance notice, the
landlord may dispose of the interest in the 12-month period from the end
of the period of nomination for a sum not less than the sum specified in
the offer notice. In the event of disagreement on the price to be paid, the
question is to be determined by reference to a rent assessment
committee.

9.16.2   Compulsory acquisition

Where a landlord is in breach of the covenant to repair, the qualifying
tenants or a majority of them may serve a notice on the landlord notifying
them of their intention to proceed under the Act for an order, giving them
a reasonable time to remedy the breach (the court may dispense with a
notice where it would not be reasonably practicable to serve one). The
tenants may then apply for an acquisition order if the landlord fails to
carry out the remedial work. For the court to make an order it must be
shown that:

- the landlord is in breach of his obligation under the lease relating to
  repair, maintenance, insurance or management of the premises or any
  part, that he is likely to continue and that the appointment of a manager
  would not be an adequate remedy or
- at the date of making of the application and for three years immediately
  preceding that date, a manager had held an appointment.

The order will nominate the person or group to acquire the interest which
will be purchased at the open market value determined by a rent
assessment committee.

9.16.3   Service charges

Where a long lease fails to make satisfactory provision for various
matters of repair, maintenance, services, insurance, recovery of expenditure on items of shared responsibility or for the computation of charges,
the court may make a variation order binding not only the parties but also

third parties including sureties. The Act extends the rights of recognised tenants' associations in relation to the employment of a managing agent and to be notified of any change in the agent or their duties.

Where service charges are payable by the tenants, they are given protection from charges levied by the landlord acting as managing agent or surveyor where it is shown that his charges are not fair and reasonable; any provision that the certificate of a surveyor is final cannot be enforced.

The 1985 Act set out the framework for landlords to provide costs and estimates for any work proposed where the cost exceeded a prescribed amount – £25 times the number of units or £500 whichever is the greater with the risk of non-recovery of any larger amounts where the landlord does not comply.

Where works are to be carried out at a cost in excess of £25 for each flat in the building or £500 in total, whichever is the greater, the landlord must obtain two estimates, one from someone unconnected with the landlord, and consult each tenant. Consultation may be directly with each tenant or by display of information in the building, in each case with copies of the estimates. Where there is a recognised tenants' association, similar information must be supplied to the secretary. Work must not be started within one month of notice to the tenants unless it is urgent. Where a court is satisfied that a landlord acted reasonably, it may dispense with some or all of the requirements for consultation.

A tenant is empowered to request a written summary of costs from the landlord with provisions as to time and certification. Any breach of the provision may incur a fine not exceeding £500, in addition to which the expenditure may not be recoverable.

The report of an action for forfeiture of the lease of a flat on the grounds of non-payment of rent and servicing charge makes interesting reading (*Woodtrek Ltd* v. *Jezek* (1982)). The case proceeded by reference to the 1972 and 1974 Acts, the provisions of both of which have now been replaced by the Housing Act 1980.

These changes affect only the details of the judgement, not its substance. The major part of the judgement concerns the landlords' claim for payment of an interim service charge and the attempt to obtain forfeiture on non-payment. A review of the then current legislation left the judge in no doubt as to the futility of the claim in the absence of receipts or other substantiation of the amount due. As far as arrears of rent were concerned, the judge held that, although the proceedings were technically justified, the behaviour of the landlords left a great deal to be desired and he granted relief from forfeiture.

The provisions do not apply to local authorities, development corporations, county councils or certain public bodies unless the tenancy is a long tenancy, when modified provisions apply. Whilst provisions in a lease cannot override the statutory provisions, it may be that the lease itself creates additional restrictions on the landlord; for example where the lease provides only for the expenditure already incurred to be recovered or restricts the periodic amount to be recovered in advance.

9.17 LONG TENANCIES AT LOW RENTS

The Rent Acts do not extend to long tenancies at low rents. For this purpose a long tenancy is defined as one for more than 21 years and which cannot be terminated before that time.

A low rent is one less than two thirds of the rateable value, payments for maintenance, repairs, rates, insurance and services to be excluded. In comparing the rent with the rateable value where the rent is progressive, the comparison is to be with the maximum rent payable. Such tenancies enabled the landlord to obtain possession and to claim for items of disrepair until Part I of the Landlord and Tenant Act 1954 gave the occupying tenant some protection.

Where the tenancy was entered into on or after 1 April 1990, the reference to rateable values is replaced by the stipulation of rents paid: not exceeding £1000 in Greater London and £250 elsewhere. Detailed provisions require that any notice served by the landlord should contain proposals for the terms to be incorporated in a regulated tenancy or give information as to the grounds on which they propose to rely in an application to the court for an order for possession.

Where a tenant has not complied with the repairing covenants of his lease and agreement is not reached between the parties, the court will intervene to determine the repairs to be carried out by the tenant. Should the landlord take proceedings more than seven months before the end of the term, the tenant may comply with any order made or elect to treat the term as coming to an end. In the latter case his liability will be limited to the payment of costs.

The Leasehold Reform Act 1967 introduced a much more radical solution by enabling an occupying tenant of a house (but not a flat) within certain rateable value limits to serve notice to acquire the freehold interest or to extend for a further 50 years subject to payment of a 'modern' ground rent. In turn, the Leasehold Reform, Housing and Urban Development Act 1993 extended the right of enfranchisement to long leaseholders of flats together with the right to obtain the right to a new lease as an alternative. Whereas the 1967 Act prescribed rateable value limits and required occupation as one of the qualifications to serve a notice to enfranchise or gain an extended lease, the 1993 Act stipulates neither and empowers occupiers of flats previously without protection from substantial rent increases at the end of their leases.

FURTHER READING

Driscoll, James (1989) *A Guide to the Housing Act 1988*, Fourmat Publishing.
Smith, P.F. (1993) *The Law of Landlord and Tenant*, 4th edn, Butterworths, London.

Index